ENERGY EFFICIENCY IN HOUSING

To our wives
Margaret, Lai-fong and Jo

Energy Efficiency in Housing

MALCOLM BELL
ROBERT LOWE
PETER ROBERTS

Avebury

Aldershot • Brookfield USA • Hong Kong • Singapore • Sydney

Published by
Avebury
Ashgate Publishing Limited
Gower House
Croft Road
Aldershot
Hants GU11 3HR
England

Ashgate Publishing Company
Old Post Road
Brookfield
Vermont 05036
USA

British Library Cataloguing in Publication Data

Bell, Malcolm
     Energy efficiency in housing. – (Urban and regional
     planning and development)
     1. Dwellings – Energy conservation   2. Architecture and energy
     conservation
     I. Title   II. Lowe, R. (Robert), 1955–   III. Roberts, Peter
     728'.04672

     ISBN 1 85972 348 9

Library of Congress Catalog Card Number: 96-83270

Printed and bound by Athenaeum Press, Ltd.,
Gateshead, Tyne & Wear.

Contents

Copyright acknowledgements

Preface

This book is based on a literature review that was carried out for the Joseph Rowntree Foundation in order to inform their own research into energy and housing. The literature review was completed late in 1994. We have taken the opportunity offered by the need to prepare the present volume, to rework much of the material in the original review, and to add two additional chapters dealing with practical experience of energy efficiency in housing. The first of these is based on work on existing housing undertaken on behalf of York City Council between 1991 and 1994. The second is based on involvement in the design and evaluation of the Longwood House, a new energy efficient dwelling in East Pennines area.

The authors of the original review were Malcolm Bell, Robert Lowe, Peter Roberts and David Johnston. The latter's contribution was invaluable, and is gladly acknowledged. A major contribution to the original literature review was made by the members of our review panel, Ian Cooper of Eclipse Research Associates, Susan Owens of the Department of Geography at Cambridge University, Anne Salvage of the Policy Studies Institute, Robert Vale of the Department of Architecture at Nottingham University, and Don Ward of BRECSU. Additional contributions were made by Stephen Edwards of Neighbourhood Energy Action, Stirling Howieson of Strathclyde University, Bill Edrich of Leeds City Council Energy Advice Unit and Mike Berrington at the Brunswick Library, Leeds Metropolitan University. The original review would have been impossible without the support of Julie Brewerton of the Joseph Rowntree Foundation.

The present authors have been assisted by Steve Curwell who contributed to the writing of chapter 9. In this context we must also acknowledge Steven Slator and William Butcher who designed, financed and built the Longwood House, and its owners, Mr and Mrs Gregory who have provided the raw data for the evaluation presented in chapter 9. In a similar vein, the work at York would have been impossible without the support of Bob Towner and his colleagues.

We are grateful for discussions with David Olivier and Chiu Lai-fong which contributed both to the original literature review and to the present book. We wish to acknowledge the care and effort expended by Janet Young in the arduous task of proof reading the final copy. Finally we must thank our families for their forbearance in the face of our prolonged absences during the preparation of this book. Any remaining omissions, errors of fact and interpretation are of course, our responsibility.

Malcolm Bell, Robert Lowe and Peter Roberts
December 1995.

1 Introduction

Over the last 25 years there has been a growing concern with energy issues. The global consumption of energy in the form of fossil fuels has now become one of the most important issues for governments around the world. The industrialised nations in particular have an important role to play in reducing consumption.

During the last 10 years the focus of concern has shifted from the problem of resource depletion and security of supply (although this issue is still important) to the problem of climate change. At the Rio summit in 1992 the UK government signed the United Nations Framework Convention on Climate Change. Among other things this convention commits the UK to taking measures which will return emissions of greenhouse gases to 1990 levels by the year 2000. Following ratification of the convention in December 1993, a programme of action was published in January 1994 (DOE et al. 1994a). The government programme is seeking to achieve a reduction of 4 million tonnes of carbon in the domestic sector by the year 2000. This represents some 40 per cent of the target for reductions in all sectors. The government actions set out to-date include: VAT on fuel (such a controversial policy that it has only been partially implemented); developing the work of the Energy Saving Trust; providing advice and information; eco-labelling; establishing standards for appliances and revision of the Building Regulations to improve energy efficiency in new building. The Rio convention and subsequent action by the UK government is an important start but it is likely that further reductions will be necessary well into the next century if global warming is to be maintained at a manageable level.

A number of studies which have dealt with the problem of stabilising the concentrations of carbon dioxide in the atmosphere, have concluded that overall global reductions of 50 per cent to 80 per cent in emission rates may be necessary during the next century. It also seems likely that reductions in the industrialised countries will need to be greater than this in order to achieve the required overall reductions while allowing developing countries to increase their use of fossil fuels in the medium term. Such reductions will be technically demanding, and their achievement will require the development of a long term strategy. Because dwellings are among the longest lived parts of our technological infrastructure, it is important that any long term strategy should pay particular attention to their energy efficiency and energy related performance.

One of the difficulties in seeking to advocate energy efficiency is that it becomes confused with the sort of deprivational conservation approaches which saw a particular vogue during the 1970s. The viewpoint of this book, however, is that energy consumption can be reduced very significantly through improvements in the efficiency with which energy is used and that it should be possible to achieve an acceptable life style and level of comfort while at the same time reducing energy consumption to a sustainable level. A deprivational conservation policy is in any case unlikely to succeed in reducing consumption without draconian measures. This is not to say that life style issues should not be addressed; there are important cultural and social differences in the use of energy which perhaps relate more to ephemeral images developed through marketing than with quality of life. One is also reminded of the work of Ivan Illich (Illich 1974) which deals in a radical way with some important issues of equity in the use of energy, particularly in relation to energy consumption differentials between the developed and developing countries. In our view, the challenge of the next century is to improve comfort levels and to allow developing countries to share the world's resources in such a way as to be sustainable in the long term.

It is of course recognised that achieving this goal will involve more than efficiency improvements. In particular, the development of renewable energy sources will be of critical importance. However, the development of renewables is unlikely to remove the need for energy efficiency. High costs and limited energy densities (renewable sources are, by their very nature, diffuse), mean that while it is technically feasible for renewables to supply most or all of the energy requirements of energy efficient economies, they can make little impact against a background of growing energy consumption.

The importance of the domestic sector is evident from energy and carbon dioxide emission statistics. In 1992 some 29 per cent of the UK energy consumption was attributable to the domestic sector (DTI, 1993), as was 28 per cent of UK carbon dioxide production (DOE, 1994a). The fact that the government's action plan, under its Rio commitment, expects the domestic sector to produce 40 per cent of the carbon dioxide savings target, further

2

underlines the importance of this sector. The potential in all domestic end use categories for energy and carbon dioxide reductions is large. Even conservative estimates based on well established technology put the overall figure at 25 per cent to 35 per cent (Shorrock and Henderson, 1990). More detailed analyses suggest that the best currently available technology may yield reductions of 45-75 per cent in individual end use categories and that if efficiency is combined with even a modest use of solar energy, an overall reduction in non-renewable energy of up to 90 per cent is technically feasible (Feist 1994).

The technology of energy efficiency is well understood, in that in most areas, technical solutions are available that are known to produce very high levels of energy efficiency. UK house designs from the late 1970's have measured space heat demands which are a half or lower than that of houses built to the current building regulations. Such designs result in houses that are warmer and use around half the energy delivered to the average British dwelling for all purposes. Recent work has produced houses which approach zero space heating requirements in the UK climate. We discuss one case (chapter 9) in which the total energy consumption for all uses is over 70 per cent lower than that of the average British dwelling. In the area of existing dwellings, recent work (one example of which is discussed in chapter 8) suggests that a halving of total energy consumption and carbon dioxide emissions is feasible, with the application of well established and cost effective methods.

It is not only in the area of space and water heating that improvements are possible. A large amount of energy is consumed in housing by the use of lights and appliances. In this sector, the literature shows scope for very large reduction in energy use (as much as a factor of 4). It is clear that even with moderate insulation measures, heating and hot water energy would no longer dominate domestic consumption and that consumption of electricity for lights and appliances would account for the majority of household energy expenditure and carbon dioxide emissions. It appears from US experience that government has an essential role to play in addressing this issue, through the adoption and enforcement of minimum appliance efficiency standards, and through energy labelling.

Although any improvement in energy efficiency in housing will require technical solutions, this is only one aspect of what is a multifaceted problem. In addition to technical factors, the overall energy efficiency of the housing system will depend on the way in which dwellings are used, the urban, suburban or rural space in which they coexist and the interfaces with other systems. For example, if as a result of improved energy efficiency, the temperature of an elderly person's home is maintained at a comfortable level,

3

their need for medical treatment may well be reduced with consequent savings of energy (and other resources) in the health sector. Although the mechanisms are undoubtedly complex (as we discuss in chapter 6) there is very strong evidence to suggest that there is a clear relationship between energy efficiency and health. Even without environmental concerns, the fact that at least 30 per cent of United Kingdom households are not able to afford to keep their houses warm in the winter and are prone to health problems related to cold and damp living conditions, provides a strong set of reasons for investment in energy efficiency. Improvements of the scale outlined above, would go a long way to reduce the high levels of expenditure needed to maintain warmth in much of the existing housing stock.

The impact of improved warmth on the provision of health services is an area which is much talked about, but as yet no clear evidence is available of the nature and extent of such impacts. However even a small reduction in doctors' visits and the need for hospital treatment would produce important savings for the health service. Similar arguments could be advanced for links between improved energy efficiency (which contributes to a more general improvement in housing conditions) and education. This is not to suggest however that energy efficiency is the only variable involved, but it has a strong claim to be considered.

If there are strong environmental, economic, social and health reasons for improving the efficiency of the housing system, how is this to be achieved? We have already observed that the problem is a multifaceted one and tackling it will require a holistic approach which seeks to understand it in all its complexity. Housing is only part of a much larger sphere of energy consumption. It would be pointless to reduce energy in the home if that meant increasing energy consumption elsewhere in the system (say in the transport sector) or if it meant missing opportunities to take a combined approach which would achieve a greater overall reduction in energy. Issues relating to housing location, neighbourhood infrastructure and the management of urban systems are important aspects in improving energy efficiency. Planning policies and development guidelines are important tools of public policy in this area. The roles and responsibilities of local authorities towards energy efficiency are being extended in a number of ways. For example, the recent Home Energy Conservation Act (1995) places a requirement on housing authorities to develop energy conservation strategies for the whole of their housing stock. Similarly, in planning, the environmental policies developed as part of the Local Agenda 21 initiative cannot be fully effective unless they take into account energy policies in housing.

Investment in energy efficiency will be an important factor in any strategies which are developed. This will require an understanding of ownership and owners' perceptions of energy efficiency. Almost 67 per cent of dwellings in Great Britain are owner-occupied and 31 per cent rented, with 21 per cent renting from a local authority; 3 per cent from a housing association and 7 per cent from a private landlord (Chell and Hutchinson 1993 - after 1991 census). Unless we understand the motivation of owners (owner-occupiers and landlords) to invest in energy efficiency and are able to devise the means by which they can be encouraged to do so, it is unlikely that the problems which give rise to energy concerns (the environment, fuel poverty, health) will be solved. In public sector housing, the Green House Programme (DOE 1993b) has demonstrated some useful results although there is still detailed work to do on policy formulation and implementation by local authorities. Further work is also required in order to extend our understanding of the nature of the indirect (maintenance and management) benefits to landlords and the extent to which they could also be demonstrated in the private rented sector. Without some form of benefit definition for private landlords, their incentive to invest is likely to be very small; so small that legislative solutions may be the only course of action if major improvements are to be made in this sector.

Although the owner-occupied sector would appear to be the most energy efficient, the absolute level of efficiency is low. Part of the problem seems to lie in the marketability of energy efficiency. In order to improve this situation attempts are being made, within the European Community, to provide information through labelling schemes for homes, the objective of which is to improve the transparency of energy efficiency at the point of sale (EEC 1993). The problem remains however that energy efficiency is a rather 'woolly' concept in the minds of many people and the link between a feeling that there is a problem and personal action is tenuous to say the least (Hedges 1991). 'Hassle' and inertia factors appear to be important and policies are required which attempt to remove some of the barriers to investment which exist.

The role of professionals in housing and elsewhere will also be critical to the removal of investment barriers. Capital cost is often seen by both clients and professional advisors as being more important than revenue savings, unless pay back times are very short (less than 5 years). Such attitudes will shape many investment decisions despite a rather more general desire to be 'environmentally friendly'.

In addition to the professionals directly involved in the management and design of housing, there are many other groups and organisations which act as intermediaries. Intermediaries provide energy advice, design and sell energy equipment, install insulation and so on. The influence of such intermediaries is complex and poorly understood but they can have a significant effect on the decisions made by householders and organisations such as local authorities

5

and housing associations. What is more, their effect often lies undiscovered. For example, it has been known for many years that the oversizing of conventional gas boilers leads to lower seasonal efficiencies, yet oversizing still takes place. There are in fact good contractual reasons for this, in that a contractor, not wanting to be faced with a complaint that the heating boiler is not warming the house quickly enough, will tend to oversize just to be safe. The important point in this context, is that unless the design is checked, the effect of the contractor's decision will usually go unnoticed and become a hidden cost to be paid by the householder in both higher fuel bills and (in the case of an owner-occupier) a higher purchase cost for the larger boiler. The incentives for intermediaries to promote energy efficiency are often weak or non-existent and as a result energy efficiency is likely to be traded off against other, more immediate, goals.

The social and behavioural dimension is an important one in both investment decision making and in the efficient use of energy by individuals in the home. Although a reasonable amount of work has been done in this area the mechanisms involved are far from clear. The overall picture is one in which, despite a favourable attitude, there are many behavioural and social barriers to energy efficiency as well as the more obvious practical (contextual) ones. On the whole; people are unsure what to do, given their individual circumstances; find it easier to adjust to price rises rather than take any risks with investment; are concerned with their levels of comfort and often do not get the best out of heating systems because the systems themselves can be difficult to understand. In addition continuing uncertainty over such things as employment further reduce an individual's willingness to engage in long term capital investment in the housing sector. The influence of social systems on the consumption of energy and the dissemination of information should not be underestimated. People would appear to build their own models of energy consumption ('Folk Models' (Kempton and Montgomery 1982)) which may be far from optimal in energy terms. These models of energy efficiency and consumption are very powerful in shaping actions and need to be understood if policies are to be successful.

This book is intentionally wide ranging. Addressing energy efficiency issues in housing will require understanding across a broad front. No single area should be seen as more important than another. Without a clear understanding of the environmental, social and economic need for energy efficiency we are unlikely to place it sufficiently high on the housing agenda. Without technology and further technological advance, the tools we need will not be available. Without a coordinated strategy, we are in danger of causing problems elsewhere in the energy system. Without an understanding of social and behavioural issues, many of the benefits of greater technological efficiency may be lost. Without a clear assessment of the resources required to develop and implement policy, very little success can be expected. Our aim in writing this book is to contribute to the housing and energy debate in a way which stresses the need for a holistic approach to the development and implementation of policy at all levels.

2 Energy, climate change and housing

Introduction: the political context

Over the last two decades the industrialised world has displayed an enduring interest in energy issues. The basis for this interest has however undergone a series of shifts. Interest in the 1960's and early 1970's was primarily based on concerns over the exhaustion of fossil fuel resources. This problem was seen as relatively long term, and was addressed by the development of nuclear power. The Middle East wars of the 1960's and 1970's and the OPEC oil embargo of 1973/4, led to a very sharp focus on oil consumption and security of supply, with a very much shorter time scale. These concerns lead to an emphasis on energy conservation measures (reduction in internal temperatures, reduced driving speeds, car free days), fuel switching (from oil to coal in non-transport sectors), and energy efficiency measures (building insulation, transport related measures including mandatory fleet performance targets for the car industry). Security of supply is a continuing concern, particularly for those countries that import a major fraction of their energy requirements. It has been reinforced by the Iranian revolution of 1979, the Iran-Iraq war, and the Iraqi invasion of Kuwait in 1991, as well as by the seemingly intractable dispute between Israel and her neighbours. The UK has been relatively insulated from the direct effects of a possible oil embargo since the late 1970's because of the development of North Sea oil and natural gas. It is however not insulated from the price effects of such events. The decline of North Sea oil production may lead to a renewed interest in security of oil supply in the UK over the remainder of the decade.

Climate change

The last 10 years have seen the rise of the third nexus of concern, which is increasingly dominant in energy policy in the industrialised nations. This is the possibility that emissions particularly of carbon dioxide from the combustion of fossil fuel will change the climate of the planet. Most analysis predicts that the average temperature of the surface of the planet will rise, and this has lead to the use of the term *Global Warming* to describe the problem. Detailed analysis shows a complex and uncertain picture of change, which has lead to the more recent adoption of the term *Climate Change*. The possibility of anthropogenic climate change through the combustion of fossil fuel was suggested in the 19th century by Arrhenius among others (there sometimes seems to be a competition on the part of late twentieth century authors to discover the earliest possible reference to anthropogenic climate change). A number of papers on climate change were published in the early 1980's (for example Bach 1980, Hansen 1981, Lovins et al. 1981). A report prepared for the US president in 1980 raised the prospect of sea level rise resulting from global warming, forcing the abandonment of coastal cities (President's Council 1980). The first major international report on the subject was published by Bolin et al. (1986). This was followed by an international conference in Toronto in 1988, at which a 50 per cent long term reduction in carbon dioxide emissions from fossil fuel combustion was proposed. 1988 saw the establishment by the United Nations and the World Meteorological Organisation of the Intergovernmental Panel on Climate Change (IPCC) under the chairmanship of Bert Bolin. The first Working Group of this panel published reports in 1990 and 1992 (Houghton et al. 1990 and 1992), and has formed the basis for much of the discussion of energy policy since then. The IPCC currently predicts a rate of temperature rise of 0.2 - 0.5 °C/decade, over the next century, under a business-as-usual emissions scenario.

It is important to realise that anthropogenic climate change is still imperfectly understood, and is likely to remain so for some time. This uncertainty is inherent in the subject matter. The earth's climatic system is very large, complex, non-linear, heavily fed-back and chaotic,[1,2] and is therefore difficult to model reliably. Uncertainties also arise because of the difficulty of predicting energy use and the direct impacts of human activity on the biosphere over the next century. Estimates of carbon emissions over the next century published by the IPCC, range from just under 700 to just under 1400 gigatonnes[3]. Figure 2.1 shows three IPCC business-as-usual scenarios. Scenario IS92a is based on medium estimates of population, economic growth and fossil fuel availability. Scenario IS92c assumes medium-to-low, and IS92e assumes high values of these three key variables. Scenario IS92e also assumes that nuclear energy is phased out by 2075. All scenarios include emissions from forestry and agriculture, which are of the order of 20 per cent of fossil carbon emissions in the first decade, but which amount to between 4 and 10 per cent of fossil carbon emissions over the whole period from 1990 to 2100. Other influential organisations have published scenarios involving carbon release both above and below this range.

The size of the uncertainty with respect to the climatic system can be illustrated by the fact that general circulation models currently in use predict that a doubling of atmospheric carbon dioxide concentration would raise global average temperature by between 1.5 and 4.5°C (Houghton et al. 1990 and 1992). Alternative hypotheses, in which changes in climate fall below the IPCC range, or in which such changes as do occur are beneficial rather than harmful, are also intellectually tenable.

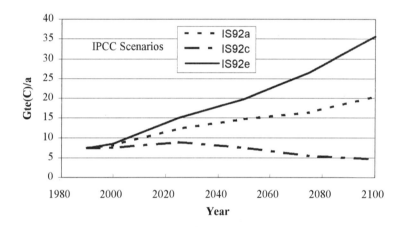

Figure 2.1   IPCC Carbon emission scenarios
Source: Houghton et al. 1992.

Measuring sustainability

The global carbon budget for the decade 1980-1989 presented by the IPCC, is summarised in Table 2.1. This table displays a mixture of relative certainty in some areas and considerable uncertainty in others. Net emissions from biomass are known very imprecisely; the IPCC estimates cover a range of over 4 to 1. Uptake by the ocean is somewhat better understood, but the estimates still cover a range of over 3 to 1. Emissions from fossil fuels, and accumulation of carbon in the atmosphere are however rather precisely known. The figure for the net imbalance is simply the sum of the four other carbon fluxes in the table, and may or may not be a sign that a significant carbon sink is still to be identified. It is important to recognise that despite the uncertainties, the quantity most directly related to climate change, atmospheric carbon dioxide concentration, is easily measured and accurately known.

Table 2.1
Anthropogenic carbon budget Gte(C)/a

emissions from fossil fuel combustion	5.4	±0.5
emissions from deforestation and land use	1.6	±1.0
accumulation in the atmosphere	3.4	±0.2
uptake by the ocean	2.0	±0.8
net imbalance	1.6	±1.4

Source: Houghton et al. 1990.

The conclusion drawn by the IPCC from this and related work, was that stabilisation of atmospheric carbon dioxide concentration at its present level could be achieved by an immediate cut in emissions from combustion of fossil fuels of around 3.4 GTe(C)/a. The precise figure depends on the details of the carbon cycle model used, but amounts to more than 60 per cent of the current fossil carbon emission rate. The uncertainty in the required cut is small - around 0.2 GTe(C)/a. Consideration of the likely effect on oceanic absorption of an abrupt stabilisation of atmospheric carbon dioxide concentration following two centuries of roughly exponential increase, suggests that following such a step, anthropogenic carbon emissions would need to continue to fall, approximately exponentially, over the following century. Emissions by the middle of the next century would need to be below 20 per cent of the current rate.

These considerations enable us to measure the degree of sustainability of present and possible future economies. It is clear that to be sustainable in the long term, human activities must not change atmospheric carbon dioxide concentration. A sustainable rate of carbon emission is therefore one which is balanced by carbon absorption by so-called 'sinks'. As an exploratory exercise, one can define a sustainability coefficient as follows:

$$S = \frac{\text{fossil carbon emissions rate}}{\text{sustainable emission rate}}$$

A world which is sustainable with respect to the carbon cycle will have a sustainability coefficient of unity. When the value of the coefficient exceeds one, the world is in an unsustainable state. Reduction of atmospheric carbon dioxide concentration requires a sustainability coefficient which is temporarily less than one. Values of the sustainability coefficient in the future can be estimated in different ways. A weak definition would equate the sustainable emission rate with the net sink magnitude in any given year, and on this basis the world mean sustainability coefficient would be roughly 2.7. However, if one assumes continued growth in emissions, and a constant air fraction[4], then the sustainability coefficient, thus defined, will not change in future years, even though carbon dioxide builds up in the atmosphere at an ever increasing rate. A stronger definition of sustainability coefficient, which avoids this apparent problem, would identify a particular atmospheric concentration as the maximum sustainable, and would use this to define a sustainable emission rate trajectory for use in the denominator of the above expression. For any such concentration, the equivalent sustainable emission trajectory will tend to decline with time from the point at which the atmospheric concentration is stabilised, by roughly a factor of 3 over the ensuing century. If one assumes that the current atmospheric concentration is the maximum sustainable, then the sustainable emission rate will fall from the current value of approximately 2 GTe(C)/a to approximately 0.7 GTe(C)/a by 2100 (Houghton et al. 1990). The resulting sustainability coefficients of the IPCC emission scenarios presented earlier, are shown in figure 2.2 below.

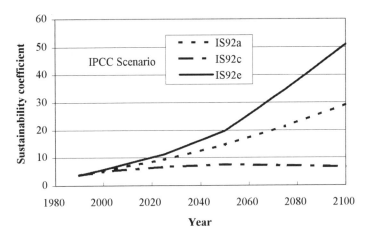

Figure 2.2   Sustainability coefficients for IPCC carbon emission scenarios
Source: Houghton et al. 1992.

The concept of the sustainability coefficient requires careful interpretation. It is clear that it is difficult to define such a coefficient in the first place. Moreover the fact that the coefficient exceeds unity is no guide either to the magnitude of the costs that may result from running the world in an unsustainable way, nor to the urgency or otherwise of returning the world to a sustainable state. A sustainability coefficient significantly greater than unity means, simply, that the activity described cannot be continued indefinitely, and will eventually diminish or cease. In this sense it is clear, that on the basis of the scenarios illustrated in figure 2.2, the world is a long way from sustainability with respect to the carbon cycle, and that the situation is as likely to get worse as better over the next century[5].

Policy responses

The arguments presented above suggest that global environmental policy needs to be made against a background of:

- the fact that human activities are currently injecting carbon dioxide into the atmosphere at a rate several times that which can be absorbed by the oceans and other sinks, with the result that the concentration of the gas in the atmosphere is increasing at an accelerating rate;

- a wide range of uncertainty over the effects of this increased concentration on the global climate, but with a possibility of changes in climate of sufficient rate and magnitude as to be harmful to part or all of the world' s human population.

The need to frame a principle that would justify potentially far-reaching action against such a background has led to the 'precautionary principle'. The form adopted by the signatories to the Rio Declaration is as follows:

Where there are threats of serious or irreversible damage, lack of full scientific certainty shall not be used as a reason for postponing cost-effective measures to prevent environmental degradation. (UN 1992)

It should be noted that while this declaration may give the impression that it commits the world to preventative action, careful reading suggests that it does not commit any party to do anything on the grounds of climate stabilisation, that they would not wish to do on other grounds.

It is extremely unlikely that the world will agree to hold atmospheric carbon dioxide concentration at or close to the 1990 value. It is however widely accepted that the precautionary principle justifies measures to slow the increase in atmospheric concentration of carbon dioxide. One of the most important outcomes of the Rio de Janeiro Conference on Environment and Development (UN 1992) were Agenda 21 and the Framework Convention on Climate Change. Article 2 of the latter states:

> The ultimate objective of this Convention [ ] is to achieve stabilisation of greenhouse gas concentrations at a level that would prevent dangerous anthropogenic interference with the climate system. (UN 1993b)

Having accepted stabilisation as a goal, the most important question becomes, at what level? As a first step to stabilisation, the Framework Convention commits the signatories to stabilising carbon emission rates at 1990 levels by the year 2000, but it is clear from the analysis presented above that this is insufficient to stabilise atmospheric concentrations of carbon dioxide.

Answers to this question depend on a number of factors and judgements, including the amount and rate of warming judged to be dangerous, the sensitivity of the climate to the changing level of carbon dioxide, and the perceived costs of reducing carbon dioxide emissions. Although profound uncertainty exists in all of these areas, a number of studies by individual nations have attempted to address the problem.

National studies

One of the most thorough and thought provoking of these was produced by the Enquete Kommission of the Deutscher Bundestag (Bundestag 1991). This considered the joint problems of world economic development, population growth and carbon emissions. It concluded that a reduction of 30 per cent in emissions rates would be needed from the Federal Republic of Germany by 2005, and that the industrialised countries as a whole would need to make reductions of the order of 80 per cent by 2050, in order to allow for growing use of fossil fuels by the developing countries within an overall target of a halving of fossil carbon emissions. This is probably the most radical suggestion to have come from a national study to date, and it is worth reviewing its argument at some length.

The Enquete Kommission began with an assertion that, based on the then current scientific evidence, energy related emissions of carbon dioxide would need to be reduced by at least 50 per cent by the middle of the next century. This position is consistent with the proposal made at the Toronto Conference in 1988 (see above). The Kommission recognised that global carbon dioxide emissions would continue to grow for a limited period, but suggested that they should fall approximately linearly from 1995 to 2050. Within this overall envelope, the Kommission then proposed that developing countries should restrict the growth in their emissions to 50 per cent, over the period to 2030, and to perhaps 75 per cent over the period to 2050. This proposal was based on a number of considerations, including the balance of existing emission rates (industrialised countries currently emit about 70 per cent of total fossil carbon), and of population (world population is currently over 5.5 billion, of whom some 1.2 billion live in industrialised countries). The remainder of the carbon dioxide emission 'rights' are then allocated to the industrialised countries. The resulting reductions required in the industrialised countries as a whole are roughly 20 per cent in 2030 and 80 per cent by 2050.

A number of observations can be made on this proposal. The emission trajectory proposed by the Enquete Kommission means that the industrialised countries would be well on the way to achieving carbon-free economies by the middle of the next century. This becomes particularly clear when one considers that the Kommission envisaged economic growth continuing in the industrialised countries over the entire period studied. The carbon emission associated with each unit of economic activity therefore falls even more dramatically than the total emission rate. It should nevertheless be noted that even on this scenario, the world as a whole does not achieve sustainability until well after 2050, using the strong definition proposed above.

A second national study of some interest is the Danish report, Energy 2000 (Danish Ministry of Energy 1990). This is of interest primarily because of its technical conclusions, which emphasise energy conservation as the primary means of controlling carbon emissions. This report presents four energy scenarios, a base scenario (business-as-usual), a supply scenario, an environment scenario and an economy scenario. Under the base scenario carbon dioxide emissions grow by about 10 per cent between 1988 and 2030, while under the latter two scenarios carbon dioxide emissions fall by 65 and 52 per cent respectively. A major conclusion of this report is that,

The analyses show that it is technically feasible to implement drastic reductions in energy consumption and the emission of carbon dioxide without this meaning greater socio-economic costs than in the base scenario if the prices of fuel rise as assumed in the projections used.

The report states that the environment scenario, as well as achieving the lowest emissions from the energy sector, also minimises foreign exchange expenditure and maximises domestic employment in the Danish economy.

In the UK there have been a number of publications from government departments or agencies which bear on the problem of energy and sustainability. These have come primarily from the Departments of Energy, Environment and Trade and Industry (the last two absorbed the first in 1991), the Energy Technology Support Unit (ETSU), and the Building Research Establishment (BRE), and range from academic studies, through informal discussion documents to white papers.

The Energy Paper series, which was started by the Department of Energy in 1975, has been continued under the restructured Departments of the Environment and of Trade and Industry. In this series, Energy Papers 58 and 59 (DEn 1989, DTI 1992) give a useful insight into official thinking. Neither paper addresses the question of how large reductions in carbon emissions from the UK economy might be achieved. Both papers continue the long tradition within the Department of Energy of basing investigations of energy policy and projections of energy use primarily upon econometric modelling, rather than on detailed technical studies of energy use.

In 1990 the UK Government published its White Paper 'This Common Inheritance' (DOE et al. 1990). This has been followed by three secondary reports. Following the signature of the Framework Convention on Climate Change in 1992, a number of additional documents have been published by various government departments. These include a discussion document on climate change (DOE 1992), a statement of the UK programme on climate change (DOE et al. 1994a), and a statement of the UK strategy for sustainable development (DOE et al. 1994b). They present a detailed and valuable picture of the present and likely future impact of the UK on the environment, under current policies. It is clear from a reading of UK government publications that while the issue of how to stabilise carbon emission rates is being considered, the problems of whether and how to stabilise atmospheric carbon concentrations is not. Should the latter ultimately become a goal of UK policy, we would expect significant changes in the political and economic environment in the UK in respect of energy related activities.

The BRE has produced two recent reports on the environmental impact of buildings (Henderson and Shorrock 1990, Shorrock and Henderson 1990). The conclusions of the second echo the conclusions of a much earlier and influential report from the BRE (BRE 1975):

A 25 to 35 per cent reduction to the carbon dioxide emission associated with the housing stock is indicated, which is sufficient in itself to reduce the overall United Kingdom carbon dioxide emission by 7-10 per cent. [ ]

The size of this estimate clearly indicates that substantial reductions to the overall carbon dioxide emission of the UK are possible through energy efficiency alone. (Shorrock and Henderson 1990)

Non-governmental policy proposals

There have been a number of non-governmental reports covering the area of energy and sustainability. One such was produced by the International Project for Sustainable Energy Paths (Krause et al. 1989), part of a continuing study funded in part by the Dutch government. The main conclusions to date are that climate stabilisation is likely to require a 70 per cent reduction in world-wide carbon emission rates by the middle of the next century, and reductions of up to 90 per cent in emissions from industrialised countries. This position is essentially the same as that taken by the Enquete Kommission (Bundestag 1991). In practice, there is likely to be little difference between the measures needed to cut carbon dioxide emissions by 80 per cent, and those needed to achieve an 90 per cent reduction. Both suggest an almost complete transition from a thermodynamically inefficient economy based on fossil carbon, to a thermodynamically efficient one based on a combination of renewable and nuclear energy.

Conclusions

Energy use in and carbon emissions from housing constitute just under 30 per cent of the UK total (DTI 1992), a fraction which is typical of many industrialised countries. If emissions remain at their present absolute level until 2050, the domestic sector alone will exceed by a factor of 1.5 the allowable emissions from the whole of the economy under the Enquete Commission scenario discussed above. In view of this, and in view of the fact that energy conservation in the housing stock is technically well understood, it is hard to escape the conclusion that large reductions in carbon emissions attributable to dwellings may be sought over the next 50 years. While in the context of practical housing programmes, it may not be possible to take account of this in the near term, we suggest that research and investment programmes and policy development should address it.

In our view, anthropogenic interference in the carbon cycle is the most serious of the sustainability issues presently facing the world, and in particular the industrialised countries. Issues other than those directly related to energy use do nevertheless exist. These include impacts of water supply, sewerage, solid waste disposal, production and disposal of building materials, and to the extent that it is affected by urban planning, food production. While there are frequently connections between these and energy related issues, we have not been able to address these issues fully in the context of the present study. This broad area has been examined by a number of authors (see for example Samuels and Prasad 1995), and is presently the focus of a joint research programme sponsored by the EPSRC and ESRC (Ekins and Cooper 1993).

Much of the material in the present volume is technical in nature. The reader should nevertheless recognise that the origins of material aspirations and needs are not technical, but psychological, social and political. For example, one of the most powerful determinants both of energy related and non-energy related environmental impacts, is the rate of new house construction. With the stabilisation of population in the UK, this is in turn determined largely by family structure and household size. With new construction rates exceeding demolition rates by a factor of 5 or 10 (Shorrock et al. 1992), 80 to 90 per cent of the current demand for new housing is generated by changing living patterns, and not by the technical obsolescence of the existing housing stock. Leach et al. (1979) suggested that average household size cannot ' fall below about 2 to 2.5 if any kind of parents-plus children family structure is to be maintained.'

Other social structures are however possible, and the two person household may not represent an ultimate limit, particularly if the ownership of second homes becomes more widespread. There may indeed be no well-defined upper limit to the demand for housing. It is clear therefore, that a transition to a sustainable economy will not be possible without a change in attitudes and aspirations of populations, particularly in the industrialised countries. How such a change is to be fostered goes beyond the scope of this book, but we tentatively suggest, particularly in view of the technical feasibility of achieving large reductions in carbon emissions from dwellings, that this question is the most important facing industrialised societies at the close of the 20th century. For further discussion of these issues we can do no better than refer the reader to the works of Nørgård and Christensen (Christensen and Nørgård 1976, Nørgård and Christensen 1995).

Notes

1. Systems such as the global climate can be considered to consist of a set of variables, the relationships between which can be described in terms of equations which encapsulate physical, chemical and biological processes. The relationship between any two variables may be considered to consist of direct and indirect effects. Systems are said to involve feedback when indirect mechanisms exist which modify the direct effect on one variable,

18

of a change in another. In the case of global climate, simple physical modelling can enable us to estimate the direct impact of increased concentrations of greenhouse gases on surface temperature. Such simple models leave out very obvious indirect mechanisms which can in principle powerfully modify the direct impact. Where an indirect mechanism reduces the size of an initial warming, a negative feedback effect is said to exist. Conversely, when the indirect mechanism magnifies the initial warming, a positive feedback effect exists. Potential positive feedback mechanisms include:

- the effect of biomass die-back due to increased temperatures, which releases additional carbon dioxide into the atmosphere following an initial build-up;

- the effect of reduced snow and ice cover due to an initial warming, which reduces the fraction of in-coming solar radiation reflected at high latitudes;

- the effect of release of methane, a potent greenhouse gas, from methane-hydrates in cold, deep oceans, and in permafrost, following warming due to an initial anthropogenic release of carbon dioxide.

Potential negative feedback mechanisms include:

- the effect of more rapid plant growth due to higher atmospheric concentrations of carbon dioxide, which reduces the initial concentration of the gas in the atmosphere;

- the effect of increased evaporation of water vapour which can result in greater cloud formation; clouds of the right type can increase the albedo (reflectivity) of the atmosphere, and thus reduce the size of an initial warming brought about by increased carbon dioxide concentration in the atmosphere.

As can be seen, plausible positive *and* negative feedback mechanisms exist in the earth's climatic system. The magnitude of many of these are poorly understood, and the overall balance is therefore hard to estimate. Feedback effects involving oceanic circulation are probably the least well understood.

2. A chaotic system is one in which it is impossible in principle to predict the system's future behaviour from its present state. Chaotic behaviour results when some or all of the relationships between the variables that describe a system are non-linear. A non-linear relationship is one in which a graph of the dependence of one variable on another is not a straight line. Many of the processes in the earth's climatic system are non-linear. Obvious examples are the evaporation and freezing of water. One would therefore expect, *a priori*, that the earth's climatic system would be chaotic. This does indeed appear to be the case, both at the level of weather (variations in a scale of hours to a few days), and at the level of long term climatic variations. The chaotic nature of weather was originally pointed out in a seminal paper by Lorenz et al. (1963).

3. It is common in climatology to quote carbon fluxes, including anthropogenic emissions of carbon dioxide, in terms of the mass flow of carbon. The abbreviation GTe(C) is used throughout for gigatonnes of carbon.

4. The air fraction is the fraction of anthropogenic carbon emissions that remain in the atmosphere, over a given period. The air fraction over the last decade has been just below 50 per cent. If emission rates continue to grow, the air fraction is likely to remain close to this level, because the rate at which the ocean absorbs carbon dioxide depends on the amount of excess carbon dioxide in the atmosphere. If emission rates fall, then the air fraction is also likely to fall.

5. One of the implications of this discussion is that the problem of climate change largely supplants the earlier concern over exhaustion of fossil fuels. This is because climate change is expected, on central assumptions, to become a serious practical problem if atmospheric carbon dioxide concentrations double, while the complete combustion of the world's resources of fossil fuels would increase atmospheric carbon dioxide concentration by a factor of 7 or more above pre-industrial levels (see for example Krause et al. 1989).

3 Domestic energy conservation technologies

Introduction

The purpose of this chapter is to review the large and growing literature on domestic energy conservation technologies, and to attempt to establish the range of options which is opened up by developments in this area. However, before beginning our review, we deemed it important to rehearse some issues relating to the taxonomy of this literature.

As with any body of literature, it is important to distinguish between primary and secondary sources. Normal practice in a scientific literature review is to present and discuss primary sources wherever possible. In the case of domestic energy conservation technology, the picture is more complex. The truth or otherwise of statements made about technology are contingent upon the broad context against which those statements are made. As the results of primary research and development are filtered through successive layers of commentary, we find these results being evaluated by qualitative or quantitative means against a variety of implicit and explicit criteria. Decisions regarding the putting in place of the actual physical infrastructure of dwellings are taken by people who are informed in the main not by primary sources, but by a variety of secondary sources, such as trade and professional journals, continuing professional development and advice from government agencies. It therefore becomes as important to review the latter as the former.

One of the characteristics of this literature is that review processes are variable, and often absent in the conventional sense. New technological facts can arise from conventional scientific endeavour, and from practice. In the former case conventional review procedures normally operate, though they may not in the case of much of the literature that results from work financed by government contract. One of the simplest requirements of any research result is that it be repeatable. The concept of repeatability is unfortunately hard to apply in important areas of technology due to the complexity of the subject matter and the economic costs involved - a prime example is the technology of passive solar space heating. Where facts arise from practice, conventional review procedures frequently do not operate. The weight that will be assigned to such facts will therefore be a matter of judgement on the part of the reader, and will vary accordingly.

Truth is contingent. Reality is, to a debatable but unavoidable extent, socially defined. The reader is urged to keep an open mind, and a well sharpened critical faculty.

Technological goals

Proposals for technological improvement can be evaluated against a number of different goals. It is possible to distinguish between the goal of incremental improvement, and longer term goals such as the desire to achieve a certain level of carbon dioxide emission.

Levels of energy demand achieved while working in the former framework are essentially arbitrary. Lower demand is generally perceived as better, but the question of whether a particular level of demand is low enough, is not meaningful. In this framework, one of the most obvious and frequently applied tests is whether a measure is micro-economically justified, under perceived present or future market conditions. This test is frequently hedged about by judgements for example as to the buildability, or marketability of a particular measure; concepts which are more problematic than microeconomic justification (see for example, London Business School 1988).

In the latter framework, it may well be possible to determine, approximately and against a background of developments in other sectors, less arbitrary targets. Questions regarding short term microeconomic justification are less important in this framework. Evaluation consists of determining which packages of measures will achieve an overall environmental target, and of determining the economic and social costs of achieving such packages.

In parenthesis, it is frequently assumed that the rate of technological innovation is limited by economic considerations - for example that insulation thicknesses in new buildings are determined by the prices of thermal insulation and energy. In other words, that what is done is what is economic. The authors' experience over the last fifteen to twenty years suggests that the opposite is equally likely to be true - that is to say, what is economic is defined by what is done.

In addition to the question of technological goals, it is useful to classify technical measures according to the degree of departure from current practice. Following a scheme introduced by Nørgård (1979), we will refer to measures as being:

moderate -   corresponding to incremental improvement of present practice, based on relatively short pay back time investments, and where considerable field experience with the measure has already been gained;

advanced -   corresponding to application of existing approaches to the limit of, or beyond what is presently microeconomically justifiable, and where field experience with the measure is limited;

radical -   corresponding to design on the basis of minimisation of full long run social and environmental cost, application of techniques for which there is little or no field experience and radical rethinking of the design problem.

Our consideration of technology is divided into sections dealing with energy demand and energy supply. There are overlaps between the two, for example in the area of passive solar design which we have chosen to discuss under the former heading.

Reducing energy demand

This section is sub-divided according to end use category - space heating, water heating, cooking and lights and appliances. In 1989 these four categories accounted for 57, 25, 7 and 11 per cent of delivered energy use respectively (Shorrock et al. 1992). These fractions have been relatively stable over the last 30 years. The single largest change has been a doubling of energy use by lights and appliances over the period. This is important because of the high resource and environmental cost of electricity compared with other forms of delivered energy.

Space heating

Space heating is the largest end use category in the domestic sector. The energy conservation potential in this sector is large - average delivered energy demand for space heating is currently between 43 and 50 GJ per dwelling. Comprehensive data on energy use in the UK domestic sector is contained in Evans and Herring (1989) and in Shorrock et al. (1992). This data is abstracted from a variety of sources including the annual Digests of UK Energy Statistics (published by the Department of Energy and Department of Trade and Industry), the quinquennial English House Condition Survey (published by the Department of Environment), the annual Housing and Construction Statistics Great Britain (published by the DOE, Scottish Development Department and Welsh Office). Energy use for space heating is not metered directly in most dwellings, and is therefore estimated from a knowledge of total delivered energy, and by a combination of modelling and sampling. Measures to improve the efficiency of space heating and to reduce total space heating demand may be investigated in a number of different ways, including large scale field trials using occupied houses, small scale trials and demonstration projects, laboratory experiments and by computer simulation.

Modelling space heat

Space heating energy use is determined in a complex way by the interplay of comfort requirements, resource availability (essentially incomes and fuel prices), external temperatures and specific heat loss of the dwelling. The physical aspects of these interactions may be modelled in number of different ways. Dynamic models take into account the energy stored over short time periods (typically 1 hour) in the fabric of the dwelling. Non-dynamic models ignore or parameterise the dynamic storage of heat in the fabric of the building, and essentially present a time averaged picture of heat flows in and from a dwelling (Clarke 1985). Dynamic models have been very extensively used to predict the utilisation of solar energy by buildings (see for example. Balcombe 1980 and 1982, Yannas 1994) and have been extensively validated (Bowman and Lomas 1985). SERI-RES, developed by Palmiter (undated) at the Solar Energy Research Institute in the 1970's and subsequently modified by Haves (1988) on behalf of ETSU, though now very old by the standards of computer software, has probably been more widely used in the UK than any other package.

In most models, energy output from the space heating system is assumed to make up the difference between gross heat loss from the building (defined as the total heat flux through the thermal envelope of the building regardless of source) and various sources of free heat gain. Space heat demand is therefore frequently referred to as "auxiliary heating".

In the UK, one of the most widely used non-dynamic approaches to modelling space heating is the Building Research Establishment Domestic Energy Model (BREDEM), developed by Uglow and Anderson (Anderson 1985). There appear to be significant differences beween predictions of space heating made by BREDEM and SERI-RES (see for example Brown 1993). The latter tends to predict considerably lower auxiliary heating requirements than the former, especially in well insulated houses. BREDEM forms the core of the two commercial domestic energy rating schemes in the UK (Chapman 1990), and of the Standard Assessment Procedure (SAP).

It is important to recognise that the benefit of any measure to reduce space heating demand or to improve the efficiency of energy supply for space heating will be partitioned between improved thermal comfort, and reduced energy use. Most of the approaches to modelling auxiliary heating requirements assume relatively well heated dwellings. Dwellings in the UK are on the whole under-heated (Evans and Herring 1989). In under heated dwellings, a relatively large proportion of the benefits of improved technology are taken as higher thermal comfort. This proportion is not easy to model, though attempts have been made (Ward 1991, Fisk 1978), and it can approach unity in low income homes (Oseland and Ward 1993). One of the consequences of this is that energy rating programs, unless used with great care, tend to over predict the absolute level of energy use and the magnitude of the savings that result from improved technology. A second consequence is that energy savings are often rather difficult to measure in housing field trials (DOE 1981), particularly where energy efficiency differences are small, and where the occupants of field trial dwellings have low incomes. One of the first field trials from the late 1970's and early 80's which did show significant experimental savings, perhaps more by luck than by judgement, avoided both of these pitfalls (Lowe et al. 1985). Detailed analysis of UK domestic energy consumption shows that while the thermal performance of the housing stock has been rising steadily over the last 25 years, delivered energy used for space heating has remained essentially static (Shorrock et al. 1992). Taking the housing stock as a whole, *all* of the benefits of energy efficiency improvements over the last 20 years have been used to raise thermal comfort standards, and energy savings have been minimal.

Reducing space heat demand

The most important factors which affect space heating requirements are thermal insulation, passive solar design, air-tightness and ventilation control, followed by heating system efficiency and control. Lowest overall energy use results from an integrated approach, in which all of these aspects are considered together.

Thermal insulation levels in new housing are regulated by the Building Regulations for England and Wales (1991, amended 1994) and for Scotland (Scottish Office 1990). Levels of insulation required by the Building Regulations were increased in 1976, 1985 and 1990. Under the latest edition of the regulations (DOE and Welsh Office 1994) which came into force in 1995, the nominal opaque elemental U values are essentially unchanged, though the calculation procedure has been improved to take account of cold bridging. The effect of the changes is to require modifications of some constructions that are allowed under the 1991 edition. The regulations also require a SAP (Standard Assessment Procedure) calculation to establish the expected overall energy performance of each design. The overall effect is likely to be a rather modest improvement in energy performance for new buildings. A recent unpublished analysis concludes that the overall performance of new houses built to current British Building Regulations is lower than that of houses built to Danish and Swedish regulations, even taking into account climatic variations (Smith et al. 1993).

The passive solar programme Much of the UK work on space heating over the last 15 years has taken place under the Passive Solar Programme. At the beginning of the period, a number of field trials of passive solar designs were commissioned, the most notable being the Pennyland and Linford Projects (Lowe et al. 1985, Everett et al. 1985). The Pennyland Project was not able to show conclusively that the passive solar measures applied, actually reduced auxiliary energy use. The main reasons for this were statistical noise and the small size of the effects being looked for, with lack of control over the experimental design being an important additional factor. The two projects established that concentrating glazing on the south side of a dwelling can yield useful savings of the order of 50-80 kWh/a/m² of glazing transferred, while auxiliary space heating demand is insensitive to total glazing area when this is added to or subtracted from a south facade. Solar savings of a few hundred kWh for the Pennyland houses were dwarfed by the difference of 11,000 kWh/a between the Pennyland houses and a nearby control group. This latter difference was ascribed to insulation improvements, reduced cold bridging, use of more efficient boilers, more air-tight construction, and a few hundred kilowatt-hours of additional passive solar gain.

The projects did establish that a combination of thermal simulation and detailed experimental work could yield valuable insights. The complexity of occupied house field trials led to the establishment of the passive solar test cell work at the Polytechnic of Central London (Ruyssevelt and Martin 1987).

This in turn became the core of ETSU's thermal simulation validation programme. In recent years, ETSU has published studies of passive solar housing estate layout (Finbow and Pickering 1988), the marketability of passive solar house designs (London Business School 1988), and has commissioned a series of passive solar house design studies (Boss et al. 1993).

In view of the comments above on the relative importance of the different groups of energy saving measures, it appears that the effort put into the passive solar programme in the UK is out of balance with research into fundamental matters such as avoiding cold bridging, and how to construct air-tight dwellings.

Government advice The Energy Efficiency Office of the Department of the Environment provides extensive information for builders and developers showing how to improve the energy performance of new housing, under the "Best Practice Programme". The aim of the Best Practice Programme is to summarise measures that are cost effective and involve minimal change to current building practice, particularly of the large volume house builders (see for example BRE 1993, Connaughton and Musannif 1990). These measures would be classed as "moderate" under our classification scheme. Houses built to the standards recommended in the Best Practice Programme are likely to have heat loss parameters in the region of 2.0 $W/m^2/K$, and air change rates above 7 ac/h, even after extensive air-tightening (see for example Palin et al. 1993). In use, these houses are likely to require more than 100 $kWh/m^2/a$ of space heating.

Examples of advanced practice The examples from the Best Practice Programme are often not as innovative or energy efficient as designs which were built and field tested in the 1970's and 1980's, for example the Salford, Linford and Pennyland houses (Webster 1987, Everett et al. 1985, Lowe et al. 1985). These designs represent an "advanced" level of performance under our classification scheme. Salford pioneered the use of very high levels of insulation and thermal mass in domestic construction, and in terms of fabric heat loss, the original Salford design still sets a challenging target for builders of new dwellings. Pennyland and Linford used somewhat lower insulation levels, but in combination with the most efficient fossil fired heating systems then available (low thermal capacity gas-fired boilers) achieved low delivered energy use and carbon dioxide emissions.

While Salford, Pennyland and Linford were field trials funded ultimately by Government, a number of individual houses have been built that come into our advanced category. These are the Longwood House in Huddersfield

(Lowe et al. 1994, described further in chapter 9), the Reyburn House (Olivier 1989), the Two Mile Ash superinsulated timber framed houses (Ruysselvelt et al. 1987), and the Charlbury Lower Watts House (Olivier 1994). Vale and Vale have been responsible for a number of houses and institutional buildings of domestic scale at this level of performance (Ashley 1988, Edwards 1990, Evans 1990). These designs use less than 50 kWh/m²/a and the best approach zero space heating demand.

Examples of radical practice A very small number of houses in the UK can be said to represent a radical approach to energy conservation. Perhaps the first were the Fielden Clegg Courtyard Houses, in Milton Keynes (Robinson and Littler 1991), which were the first houses in the UK to use a glazing system with a centre-pane U value of less than 1 W/m²/K. The best recent example is the Vales' Southwell House (Bunn 1994), which combines a number of features to produce a house which is expected to come as near as any in the UK to the goal of autonomy. We should also mention Peter Carpenter's Caer Llan field study centre (Carpenter 1990), which is probably the first domestic scale building in the UK to have achieved zero space heating demand. Carpenter's earth sheltered design appears to suffer from interstitial condensation problems, as a result of being internally insulated. It appears to have inspired Vale and Vale to produce a similar (but externally insulated) design for low income housing (Anon 1994).

Several of the houses in the advanced category make use of mechanical ventilation with heat recovery. Such systems can reduce overall energy use in the UK, but are still comparatively expensive. They also require high levels of air-tightness (Liddament 1993). Air leakage measured by pressurisation testing should be, ideally, less than 1 air change per hour (ac/h) at a test pressure of 50 pascals (Pa) and certainly less than 3 ac/h at 50 Pa. Such levels of air-tightness have been demonstrated in timber framed houses in the UK (Ruysselvelt et al. 1987, Scivyer et al. 1994), and recently in masonry superinsulated houses (Lowe et al. 1994, Olivier 1994). Average air change rates in the UK are however much higher (Perera and Parkins 1992). It is likely that in houses of intermediate air-tightness (3-5 ac/h at 50 Pa), one of several forms of extract ventilation (passive stack, mechanical or hybrid) may perform better overall. More work is required to demonstrate air-tight construction, and to produce reliable design guidance for building professionals. Achieving air-tight construction requires considerable attention to detail in both design and construction. In addition to design improvements, the ability of the construction industry to develop the skills and the quality control methods required will be critical. This may be difficult for an industry that has adopted a policy of de-skilling and casual employment over many years.

North American and European experience There is considerable experience on the mainland of Europe and in North America of construction techniques that would be classed as advanced or radical in the UK context. Houses in North America and Sweden are predominantly timber framed. Unless considerable care is taken, timber framed construction is not air-tight. This problem was addressed in the 1970's in Canada and Sweden, as a result of which, comprehensive and well tested advice now exists for the construction of superinsulated timber framed dwellings with arbitrary insulation thicknesses, and air leakages below 1 ac/h at 50 Pa (Carlsson et al. 1980, Energy Research and Development Group 1980, Nisson 1988). Canada has seen two government sponsored initiatives to develop the concept of energy efficient and low environmental impact homes; the R2000 Programme (Cole 1993), and the Advanced House Programme (Dumont 1992, Dumont 1993, Sinha and Mayo 1993), which aim to reduce delivered energy use by roughly 50 per cent and 75 per cent compared with houses built to 1975 Building Codes. The result in the case of the Waterloo Advanced House is a 234 m² dwelling that is expected to use 17 per cent less delivered energy in the climate of Ontario than a typical 90 m² Pennyland house in the climate of England (Carpenter and Kokko 1993, Grady 1993).

Recent experience of very low energy housing in continental Europe has been summarised by Olivier (1992). Of particular interest is experience in Germany and Switzerland, which have climates similar in severity to that of much of the UK. Olivier describes projects which are expected to use 5 kWh/m²/a or less of space heating - this compares with the current UK average of about 180 kWh/m²/a, and around 80 kWh/m²/a for a Pennyland house. The 95 per cent reduction is in line with requirements for stabilisation of atmospheric carbon dioxide concentration (see chapter 2). Swiss and German experience suggests that very low air change rates (around 0.2 ac/h at 50 Pa) are relatively easy to achieve in masonry construction. Much of the work on the continent has concentrated on eliminating cold bridging from constructions (Brunner and Nänni 1990, Arbeitsgemeinschaft für zeitgemässes Bauen 1989).

Advances in glazing

Over recent years, the performance of glazing systems has advanced considerably. Initial advances concentrated on reducing centre pane U values of glazing by use of low emissivity coatings, heavy gas fill, use of convection suppressing systems for gas spaces, introduction of evacuated and aerogel systems (Halcrow Gilbert Associates 1992, Lampert and Ma 1992). The result is a number of windows now on the market with centre pane U values of less than 1.0 W/m²/K compared with the U value of conventional double glazing of approximately 2.8 W/m²/K. Suppliers include Accurate Dorwin Co of Winnipeg, Canada and Geilinger AG of Switzerland (Anon 1993, Olivier 1990). In recent years there has been a realisation that losses from the edges of windows can mean that the whole window U value is more than 50 per cent worse than the centre pane U value (Lowe and Olivier, 1995). The effect of edge losses becomes larger as the centre pane U value falls. The result has been the development of insulating glazing spacers, and window framing systems that reduce edge losses (Schultz 1992, Edgetech Newsletter). The Swedish Government agency NUTEK recently staged a competition for the design of a high performance timber window. The result was a number of windows with overall U values of 0.73 - 0.9 W/m²/K (Persson 1994). In the UK such window systems would be net gainers of heat in all months of the year, if placed on a southerly facade.

The importance of high performance glazing lies in that a large proportion of window area in UK dwellings is still single glazed. Lifetimes of windows in UK dwellings are of the order of a few decades, significantly shorter in most cases than the lifetime of the whole building. There are therefore significant opportunities for upgrading existing dwellings, as windows come to the end of their useful lives. The use of the most advanced window systems, with U values of close to or below 1.0 W/m²K under these circumstances is a relatively cheap route to high thermal performance in many existing buildings.

Water heating

The principles of reducing the energy demand for water heating are the reduction of heat losses from heat generators and stores, the reduction of distribution losses, the reduction of hot water consumption at point of use, and the recovery of heat from waste hot water. We are aware of some work in these areas (see for example Evans and Herring, 1989). Much of the rather subtle analysis work from the 1980's on the efficiencies of gas fired water heating systems is now largely obsolete, since condensing gas boilers have efficiencies close to 90 per cent regardless of load, removing the distinction between marginal and average efficiency. There appears to be little literature on the subject of waste water heat recovery, and even the most advanced low energy projects have treated this as a technology of last resort (Kreisi 1989). Measures to reduce hot water consumption include the use of spray head taps and presence-sensing devices, none of which have yet been widely used in the domestic sector. It has been suggested that mains pressure showers with low-flow heads may significantly reduce the energy requirement for whole body washing, however, unless satisfactory low-flow heads are widely available and are used in preference to high-flow heads, significant savings from mains pressure showers are unlikely to materialise. There is also evidence that flow rates with conventional shower systems may have been overestimated (Nisson 1988).

Attention over the last 20 years has focused on space heating energy use. Much of the work that has been done on water heating has been in the context of active solar heating programmes (Alper et al. 1985), and has not focused on reducing demand. Energy use for water heating can easily exceed space heating in well insulated and air-tight houses. This would seem to be a fruitful area for research.

Cooking, lights and appliances

These areas of demand account for about 18 per cent of UK domestic sector delivered energy use, but over 30 per cent of carbon emissions. In well insulated houses this latter fraction can be as high as 75 per cent. Very large reductions in energy use by lights and appliances are possible. Nørgård (1979) suggested up to 65 per cent compared with the average 1975 Danish model. Although there have been subsequent improvements in energy efficiency of typical appliances, technological advances have meant that the savings potential has more than outstripped these, with the result that ten years after publishing his first estimates, Nørgård estimated that the savings potential had grown to 74 per cent of Danish electricity use for domestic appliances, at an average cost of 2.5 US cents per kWh (Nørgård 1989). These estimates are based on fundamental engineering analyses of energy use by each type of appliance. Estimates of savings for the UK based on replacement of current stocks of appliances by the best on the UK market and the best on the world market are 40 and 45 per cent respectively, with 60 per cent savings being possible through use of developments not yet on the market (March Consulting Group 1990).

There are significant barriers to the introduction of more energy efficient appliances. With respect to energy efficiency, purchasers of appliances appear to have implicit discount rates which approach 100 per cent (Williams 1989), though these are perhaps more likely to be evidence of a lack of attention to energy conservation issues (see discussion of behavioural models in chapter 7), and an unwillingness to believe manufacturers' claims, than for intrinsically pathological time horizons. Some energy efficient appliances, for

31

example compact fluorescent lamps are twenty times as expensive as the inefficient appliances they replace, and even though these are widely available, and even though they more than halve the effective cost of light, market penetration is still small. In other cases, for example refrigerators, the additional manufacturing cost of more efficient models is small, but the most efficient models (for example. Gebruder Gram's LER 200, and the Californian Sunfrost models) are not available in the UK. In the UK, energy labelling of new appliances would aid consumer and retailer awareness of energy efficiency. The EC directive on energy labelling of appliances should eventually achieve this (EC 1994).

There is evidence from the US that minimum appliance efficiency standards, introduced under presidents Reagan and Bush in 1990, have had a major effect on energy efficiency of refrigerators (Rosenfeld 1993). It may well be that such regulatory action is the most appropriate short term step to take, in areas where the nature and scope of the engineering improvements that can be made to improve performance are simple and obvious, the marginal costs low, and where the scale of the market failure is egregious. Minimum appliance standards simultaneously stimulate manufacturers to improve the performance of their appliances and ensure that consumers buy the improved models. We understand that the EC will introduce such standards in 1997, though the initial cut-off level chosen will improve average efficiency by only 10 per cent (DOE et al. 1994a).

A number of alternatives to regulation exist. Both the US and Sweden have staged competitions in which manufacturers are challenged to produce domestic appliances meeting a given level of performance. The prize (the so-called Golden Carrot) for such competitions is often in the form of a guaranteed contract for the first batch of production of the new appliance. Other approaches are exemplified by the Canadian Advanced House Programme which establishes comparatively rigorous voluntary minimum appliance efficiency standards (Dumont 1992), which challenge developers who wish to participate in the scheme, to source the most efficient appliances available. The recently published Environmental Standard for Dwellings (Prior and Bartlett 1995) sets a challenging target for carbon dioxide emissions per unit of floor area, which can only be met by dwellings which incorporate low electricity appliances, and will encourage UK developers to do the same.

Energy supply

Present position and near-term future

Primary energy input to the UK in 1990 consisted of 2.9 EJ of coal, 3.2 EJ of oil, 2.2 EJ of natural gas, and 0.8 EJ of primary electricity (DTI 1992). Total UK primary energy demand has not changed significantly since 1970, though within the total, use of coal has fallen, while use of gas has increased. The effect of the changing mix has been to reduce the carbon intensity of primary energy consumption by about 18 per cent, a trend that has continued since 1990 with the displacement of large amounts of coal from the generation of electricity.

The carbon intensity of the main energy carriers which supply the domestic sector differ widely. Natural gas has the lowest carbon intensity at 0.19 kg/kWh, followed by oil at approximately 0.27 kg/kWh, coal at approximately 0.30 kg/kWh, and electricity at approximately 0.68 kg/kWh (DOE and Welsh Office 1994)[1]. The carbon intensity of natural gas is fixed by virtue of the minimal amount of energy required to deliver methane to the consumer, and by the chemical nature of methane. The carbon intensities of the other energy carriers are to a greater or lesser extent variable, and in the case of the secondary fuels (coke, oil and especially electricity) depend on the technology of their production. In the case of electricity, changes in the mix of production, principally reductions in coal and increases in natural gas fired plant, are expected to lead to a further fall in carbon intensity by about one third from the present to the end of the decade.

Energy delivered to the domestic sector is roughly 64 per cent natural gas, 23 per cent electricity 10 per cent solid fuels, and 5 per cent oil. Consumption of solid fuels is likely to continue to fall for the foreseeable future, especially given the fact that UK production of house coal has all but ceased. Given the environmental and social and personal costs of solid fuels, this development is to be welcomed, though the authors regret the way in which it has been achieved. The next 10 years will therefore see the majority of the domestic low temperature heating load (space and water) and about two thirds of the cooking taken by natural gas, and most of the rest by electricity. Evans and Herring (1989) point out that the total size of the off-peak electricity resource is of the order of 60 PJ/a, approximately 10 per cent of total UK useful heat requirement for space heating. Off-peak electricity does not therefore represent a strategically important option for low temperature heating in more than a relatively small fraction of the UK domestic sector. Competition between the two energy carriers in the heat market will therefore be marginal in the short term.

This position could change in the longer term in a number of ways. The development of electrically driven heat recovery technologies will enable on-peak electricity to supply some of the heat market at a cost comparable to that of gas (McIntyre 1986), though the problem of peak loads imposed on the electricity generation and distribution system by electric space heating will limit applications where electric-generated heat supplies the whole of the low temperature heat load. Mechanical ventilation heat recovery using simple heat exchangers (MVHR) in gas heated houses does not add significantly to peak load. Indeed the effective coefficient of performance of such systems increases as the outside temperature falls, making them, in overall terms, rather attractive sources of space heat.

District Heating and Combined Heat and Power

The most important additional options for supplying the domestic sector over the next 20 years, are probably combined heat and power and district heating (CHP and DH) and active solar. Combined heat and power is capable of reducing the energy use by the domestic sector by about 30 per cent. The exact figure depends on many variables, including the heat-to-power ratio of the primary energy converter, the comparative thermodynamics of electricity-only generation compared with CHP generation, energy losses in heat distribution, and the comparative efficiency of heat supply from conventional sources. CHP can be considered at a number of scales. Micro-CHP is normally taken to refer to systems based on engines up to a hundred kilowatts electrical output. City scale CHP makes use of power stations of a few megawatts up to hundreds of megawatts. Essentially all scales of operation are now possible, down to a lower limit of about 10 kilowatts electrical.

The problems associated with city scale CHP were considered in the 1970's and early 1980's with the publication of Energy Papers 20, 34 and 35 (Combined heat and Power Group 1977, 1979a and 1979b). In the 1980's a number of "Lead Cities" were chosen (Atkins 1984), in at least one of which it was expected that CHP/DH would be demonstrated. Other events, particularly the privatisation of the electricity industry, intervened and the schemes envisaged by the Atkins report did not go ahead. The partial exception is Sheffield, where a heat-only scheme is now in existence.

Small scale CHP has been the subject of a number of studies, one of the most recent being by Evans (1990), who concentrates on the institutional and industrial markets for the technology. Studies of small scale CHP in the domestic sector have been done (see for example Everett and Andrews 1986), but the economics of domestic applications have worsened steadily since the privatisation of the electricity industry (Everett 1992), due to increases in the fixed portion of electricity tariffs. The most obvious and immediate role for small scale CHP is in high density housing, especially in tower blocks where natural gas cannot be used. Several such schemes have been implemented, the latest under the Greenhouse Programme of the Department of the Environment (DOE 1993b). It must be said that in at least one of these schemes, at Leeds, the CHP heat output capacity is only one tenth of the total. The CHP component was restricted to this very low level because the local authority landlord could not see a way of supplying more than his own demand for electricity in the block served. Thus CHP was restricted to supplying the lift and pump motors, and communal lighting only. The question of the tariff structures and rules under which CHP operates are central to the question of whether or not this technology will have a part to play in the UK in the future (Everett 1992). The forthcoming liberalisation of the electricity and gas markets may be the means by which the institutional barriers to CHP will be overcome, but given the history of the last twenty years, we are not hopeful.

It is useful to compare the position in the UK with that in Denmark. Following the second oil crisis in 1979, the Danish Government passed the Heat Supply Act, which initiated a process of nationwide heat supply planning. Under this act, an initial fast planning process was set up to cover the most densely built up areas (Danish Ministry of Energy 1987). At the time of the first Heat Supply Act, coverage of district heating was approximately 40 per cent of total heat load, and is now approximately 50 per cent. The 1979 act was followed by the 1990 Act, which among other things ruled that all DH plants over 1 MW be converted to natural gas fired CHP, or to biomass over an eight year period. By 2005, 60 per cent of Denmark is expected to be connected to communal heating systems, with over two thirds of these being supplied by CHP (Danish Ministry of Energy 1993). The concerns that underlay the development of DH in Denmark are security of supply, and a desire to control and reduce carbon dioxide emissions. DH is seen as a highly flexible system, which enables a wide range of energy sources, including renewable sources such as biomass, solar and wind, to be harnessed to heat supply.

It appears from our brief review of the literature that many of the most important questions in the area of DH and CHP remain institutional and political rather than technical. Research into the former must however be aware of the technical developments that are being made. The most important of these are the development of low temperature DH systems, operating at supply temperatures below 70°C at an outside temperature of -12°C, the development of low cost heat supply systems, and the integration of renewable energy into DH systems (Danish Ministry of Energy 1993, Meyer and Nørgård 1989).

Active solar

Active solar has been extensively researched in the UK and elsewhere. The UK maintained an active solar research programme between 1977 and 1982 (Alper et al. 1985). Early work on active solar space heating showed that this does not make sense in uninsulated dwellings. In new dwellings, it now seems that removing the space heating load completely is simpler and cheaper than attempting to provide a large fraction of it by active solar (Dumont 1993).

Individual domestic hot water heating appears to be a more promising role for active solar. A significant problem in the UK seems to be that overhead costs in the active solar industry are very high (Gillett and Stammers 1992). This results from the small size of the present market, but also tends to prevent the growth in the market that would drive all costs down. The Dutch Government is currently attempting to overcome this problem through a short term subsidy (Bosselaar 1994). It is important to realise that active solar is unlikely to reduce the total price of domestic water - on the most optimistic assumptions, active solar in the UK produces heat at many times the marginal cost from a gas fired heating system (Gillet and Stammer 1992). It should however reduce the marginal cost of hot water, particularly in the summer, and will clearly reduce the emissions associated with hot water heating.

Developments in glazing technology, particularly transparent insulation materials, offer the possibility of significant improvements in performance and simplification of active solar systems (see for example Kreisi 1989). The use of photovoltaics to power circulation pumps in active solar systems seems to result in significant simplifications of control systems, as well as the effective elimination of parasitic electricity requirements (Sinha and Mayo 1993). In the long term it is possible that solar heating integrated with district heating systems will offer an alternative to natural gas as the main energy carrier supplying the domestic low temperature heat market.

Photovoltaics and other renewables

Great strides are being made in the development of photovoltaics (Derrick et al. 1993), which are currently economic in many independent applications. Buildings clearly offer interesting possibilities for mounting photvoltaics close to loads (Hill et al. 1992). However most dwellings in industrialised countries are, and are likely to remain connected to electricity grids. Photovoltaics do not have a major impact on the design of buildings, and in grid-linked applications will not have a major impact on the inhabitants of houses. Grid-linked houses provide useful platforms for photovoltaics, which can relatively easily be retrofitted to most houses.

The same considerations apply to other technologies such as wind power. The best wind sites are not normally in urban areas, and in a country in which the overwhelming fraction of dwellings are grid-linked, there is no necessarily close relationship between the requirements of dwelling and those of wind power production. To the extent that these technologies are introduced, their main impact will be on the general energy supply context of the dwelling rather than on the design of the dwelling and the lifestyle of its occupants.

Conclusions

The overriding impression left by a review of the current status of domestic energy conservation technologies, is that the means exist to make large reductions in energy use in all areas of demand. A guarded optimism regarding the long term future of energy use in the domestic sector and its resulting environmental impacts is probably justified, provided that action begins soon to implement the technical options sketched above.

In our view, action is most urgent in new housing. If one takes the hubristic step of projecting current patterns of demolition and new construction forward to the middle of the next century, by that time some 40 percent of all housing would have been built since 1995. Although the construction process itself inevitably impacts on the environment, the impact of these houses once built could be a small fraction of the impact that already arises from the present stock of dwellings. Opportunities for energy conservation in housing that is still on the drawing board are very large and the marginal costs are modest.

The situation in existing housing is more complex. Significant numbers of existing dwellings are appallingly badly built, and renovation costs are high. Demolition of a significant portion of the existing housing stock may well be a better environmental option over a period of fifty years than attempting to improve their energy performance by retrofit. But even in existing housing, a carefully planned programme of improvements undertaken over a full maintenance cycle can yield significant energy savings at modest marginal cost (the reader is referred to chapter 8 for a description of experience in one local authority).

At present, measures to reduce energy demand show rather more promise than measures on the supply side. We suspect that this situation will continue for some time. The energy supply technologies that are currently in a position to make a major impact on the domestic sector are solar hot water heating and photovoltaics, and of these only the former interacts in a major way with the design of the house and its services systems. Energy conservation concerns, above all, the thermal envelope of the dwelling. Getting the best performance from the thermal envelope requires that energy conservation considerations be included from the outset of the design (or in existing housing, the redesign) process.

Energy conservation does more than reduce emissions of carbon dioxide. Perhaps its most important effect is to make dwellings more pleasant and healthier places to live in. The technical measures which would be needed to make a major reduction in carbon dioxide emissions from UK dwellings, would effectively eliminate fuel poverty, and the ill-health and human misery that it causes. Energy conservation represents, to a large extent, a no-regrets approach to avoiding the environmental impacts of energy use in the domestic sector.

The guarded optimism with which we began has, however, to be tempered with respect for the practical, political and social problems which stand in the way of implementation of technical solutions. These matters will be dealt with in the following chapters.

Notes

1 Carbon dioxide emissions may be quoted in terms of mass flows of carbon, or of carbon dioxide. Scientific work on the carbon cycle normally adopts the former convention, since although carbon appears in different chemical forms in different parts of the cycle, the carbon atom is present in all of them. We used the former convention in the preceding chapter, but have adopted the latter more widely used convention for this and subsequent chapters.

4 Structures of ownership and public policy

Introduction

Energy efficiency depends not only on the availability of technology but also on the systems and institutions through which the technology is applied. These in turn are established and developed by government through the political process. The objective of this chapter is to consider the nature and impact of some of these systems and how they have affected the development of energy efficiency in the housing sector. We conclude with a review of government policy and a discussion of possible future policy options in this area.

Structures of ownership

In Great Britain some 69 per cent of dwellings are owner occupied and approximately 20 per cent are rented from a local authority, with the remainder being rented privately or from housing associations (Chell and Hutchinson 1993 - after 1991 Census). The mechanisms which determine the energy efficiency of these different portions of the housing stock differ markedly, and the resulting levels of energy efficiency show consistent differences. Differences in tenure are reflected in and paralleled by differences in social class and household income, leading in turn to consistent differences in the household perceptions of the importance of energy efficiency in each sector.

An extensive energy survey formed part of the 1986 and 1991 English House Condition Surveys. Data from the 1991 survey are expected to be published shortly. The 1986 data demonstrate a low level of energy efficiency among rented dwellings compared with those which are owner occupied.

While 31 per cent of owner occupied dwellings were in the top efficiency quartile (classed as 'efficient homes'), only 12 per cent of council tenants and 8 per cent of private tenants had homes in this band. At the opposite end of the efficiency spectrum, 36 per cent of local authority dwellings and 51 per cent in the private rented sector were in the lowest quartile ('inefficient homes'), with only 14 per cent of owner occupied dwellings in this band. The housing association stock (2.6 per cent of the total), has large proportions of houses in both top (40 per cent) and bottom (33 per cent) energy efficiency quartiles. More detail on the energy efficiency status of various parts of the housing sector is provided in Evans and Herring 1989, Shorrock et al. 1992, Shorrock and Brown 1993, Dunster 1994a, 1994b and 1994c.

Structures of ownership and household income can explain a large portion of these differences. The most important determinants of energy efficiency of any given dwelling, are the building regulations in the year of construction, and the amount of subsequent expenditure on energy efficiency retrofit measures. Energy efficiency standards effectively entered the building regulations for the first time in 1975, and most of the dwellings built since 1980 (some three million) have been built for the private sector. Over this period, housing associations have been the main source of new social housing, largely displacing local authorities. This largely explains the relatively high proportions of 'efficient homes' in both owner occupied and housing association sectors.

Energy improvements to existing dwellings tend to affect the least efficient dwellings most strongly. The high proportions of 'inefficient homes' throughout the rented sector, compared with the very low levels in the owner-occupied sector, are almost certainly a sign of low levels of repair and maintenance in the former compared with the latter. This in turn can be explained by significant institutional barriers to investment and low levels of disposable income that exist in all parts of the rented sector.

It must be born in mind that the definitions of 'efficient' and 'inefficient' referred to above are relative. Even properties built to 1990 Building Regulations standards would not satisfy the criteria for affordable warmth for many low income families (Boardman 1993). We also recognise a number of changes that have occurred since 1991, with a much greater emphasis on energy efficiency in housing investment in the public sector. In spite of this, we suspect the picture sketched above, of energy inefficiency concentrated in the rented sector, will remain for some time to come. Moreover, we expect the differences in social class and income, outlined above, to become starker.

The owner occupied sector

As noted above, the owner occupied sector is characterised by relatively high levels of disposable income, which make it possible for households to consider making marginal investments in energy efficiency. Such investments benefit the households that make them through lower fuel bills and increased comfort, and in principle also through higher resale prices. The purpose of this section is to examine some of these issues in more detail.

Energy efficiency measures may be divided into a small number of measures that pay back extremely quickly, and a larger number of measures that have much longer pay back times. In the first category are such measures as draught-proofing, adding loft insulation up to a thickness of 150 mm, simple measures to control energy used for water heating and so on. To improve the performance of a dwelling significantly beyond the level of the current Building Regulations requires investments which have relatively long marginal pay back times. Installing condensing instead of conventional gas boilers, low emissivity inert gas-filled glazing instead of simple double glazing, loft insulation beyond 200 mm, and more than 100 mm of wall insulation, all have micro-economic pay back times which are in the region of 10-20 years.

Although the total sums involved in these investments are small compared with the total cost of the dwelling, the pay back times are longer than the average length of tenure in this sector. Thus the level of investment is determined, to a large extent, by the marketability of energy efficiency measures at the point of sale and resale.

It has been argued (Pyle and Davies 1994) that increasing the supply of energy efficient homes (above the minimum Building Regulations standard) will depend on policies designed to increase demand, but in our view any analysis that emphasises any one aspect of the problem is suspect. A large number of factors affect marketability of energy efficiency measures, whether applied to new or existing homes. These include the cost of energy compared with other household expenditures, cost effectiveness of energy efficiency measures, uncertainty regarding future energy prices, personal and family insecurity, availability of energy efficiency (in the form of services, products and efficient houses), the problem of assessing claims of energy efficiency, and ignorance on the part of householders, throughout the construction industry and among housing professionals. Some of these issues are dealt with in more detail in subsequent chapters of this book, but we wish to make a number of observations here.

Average expenditure on energy is less than 5 per cent of household income and among owner-occupiers the proportion is even lower (Shorrock and

Brown 1993). Leaving aside for a moment the problems facing low income families, for whom energy expenditure can exceed 10 per cent of a much smaller income, this is a tiny proportion.

Secondly, although energy prices excluding transport fuels rose with respect to the retail price index by some 28% per cent from 1975 to 1987, they fell over the period from 1987 to 1993 (see figure 4.1). The fall in price was briefly reversed by the partial imposition of VAT in 1994, but otherwise is likely to continue. Most of the fall in prices has been accounted for by a falling real price of natural gas, down by 23 per cent between privatisation in 1986 and 1993 (DTI 1994). The message that is conveyed to householders by this recent price history, is that energy is cheap and likely to get cheaper.

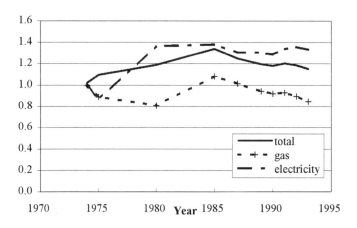

Figure 4.1   **Relative domestic energy prices** (Jan 1974 = 1)
Source: DTI 1994

Thirdly, the availability of energy efficiency in the form of services and products, and even more so in the form of energy efficient dwellings, is poor. Few builders are prepared to build dwellings which go beyond the requirements of the Building Regulations (we document an exception in chapter 10), and the small amount of evidence which is available suggests that house buyers are likely to be reluctant to pay the additional costs of a low environmental impact house (Webzell and Fewings 1994). Would-be builders of energy efficient houses frequently encounter difficulties in sourcing key products (examples from the authors' experience include long plastic wall ties, glazing incorporating warm edge technology, and insulated timber beams). High price tends to be the obverse of poor availability and low demand. One of the best documented examples of this is that of active solar water heating, referred to in chapter 3. Even where a technology is widely used, economies of scale may exist and may exert a powerful influence on uptake. A good example of this is cavity fill, which is typically three times as expensive in the context of one-off contracts in the owner-occupied sector than it is in local authority housing. Current one-off costs for cavity fill are £300 to £450 per house (EEO 1991), while contract costs in local authority housing are around £100 to £150 per house (Bell et al. 1992). Further evidence of the benefits of bulk contracts is given in a project carried out by the Bournville Village Trust in 1987. There the contract cost of £85 per house was just over a third of the cost of a typical one-off cost. When a special discounted rate was negotiated by the Trust for owner occupiers on their estate, some 400 people took advantage of the offer (BRECSU 1990). This example suggests that the costs imposed in the private sector by one-off operation can be a significant barrier to uptake of energy efficiency measures.

The problem of assessing energy efficiency claims at the point of sale have been widely appreciated for many years. Recent developments include an EC directive which among other things requires member states to draw up and implement programmes on the energy certification of buildings (EEC 1993).

The directive does not specify any particular means of implementation, and leaves this to individual states. In Denmark, a government-backed energy inspection scheme has been in place since 1975 and, since 1985, an energy report or certificate has been required for any dwelling offered for sale which was built before 1979 (Danish Ministry of Energy 1990). As a result of this and other schemes, total energy use for domestic heating per square metre of accommodation fell by around 45 per cent from the mid 1970s to 1990. Over the same period in the UK space heating consumption remained roughly static. In the UK at least three energy labelling schemes are now in operation. From July 1995 there has been a requirement under the Building Regulations for all new dwellings in England and Wales to be energy rated using the Standard Assessment Procedure (SAP). The extent to which the labelling of new and existing houses achieves greater discrimination on energy grounds is uncertain, but Hedges (1991) reports that most people interviewed welcomed the opportunity it would provide to make an informed choice.

In theory the owner occupier has the greatest incentive to invest in energy efficiency because the pay back is direct, because comfort levels in owner occupied houses are generally higher than in the rented sectors, and because owner occupiers possess disposable income which can be invested in energy conservation. Our discussion nevertheless suggests many reasons for the slow progress, documented in successive House Condition Surveys, in this sector.

Policies which might increase the rate of investment in private housing are discussed in the final part of this chapter.

The rented sector

The main problems in this sector, in addition to the impacts of energy use on the wider environment, are poverty, poor quality housing, low internal temperatures and ill-health. This syndrome of problems is widely referred to as fuel poverty, a term originally introduced by Boardman (1991a). Two major barriers to investment in energy efficiency in this sector can be identified. The first is that investment costs are normally borne by the landlord, while the most easily identified benefits accrue to the tenant. The second, is that poor households in under-heated dwellings tend to take a large fraction of the benefits of improved energy efficiency as higher comfort, rather than as a financial saving which could in principle be used to pay for investment in energy efficiency.

Even if tenants wish to invest for their own comfort and financial benefit, many are unlikely to be able to do so. Some 70 per cent of the poor live in rented accommodation (Boardman 1991a) and in local authority housing about 63 per cent of tenants claim a means tested benefit. In housing association homes the proportion is almost 67 per cent (NEA 1993a).

Many local authorities and housing associations are becoming increasingly aware of their role in the alleviation of fuel poverty but, at the same time, they face other important calls on reducing resources:

Energy efficiency has thus had to compete with other priorities, against a background in which central Government control of local authority capital investment increased significantly and the finance available declined. (AMA 1991)

The scope for sharing the financial pay back between tenant and landlord through increased rents has been explored (see for example BRECSU 1992). Such an approach is probably harder to introduce in existing housing, where it would involve raising existing rents. The propensity of low income tenants to take the benefits of energy efficiency in the form of higher temperatures, means that in many cases a rent increase would be difficult to sustain without additional financial support, perhaps through housing benefit. All studies in this area recognise that estimates of tenant savings are difficult to make, because most tenants would not have been able to afford to heat their homes to adequate standards before improvement. In one striking example from Glasgow (BRECSU 1993c), poor insulation and a particularly expensive

44

heating system combined to produce an estimate of energy expenditure in a single dwelling before renovation of £1750 per year, a figure which could exceed 30 per cent of the income of a household supported by state benefits. The unreliability of energy audit calculations have been shown by Mayers (1993) who found actual energy expenditure of as little as 43 per cent of predicted.

Indirect benefits to the landlord, which do not depend on raising rents, can nevertheless be identified in principle. A number of case studies prepared for the Energy Efficiency Office by BRECSU (BRECSU 1993a -1993d), have attempted to quantify these benefits based on information from three local authorities and one housing association. The prime sources of benefit are: reduced voids, lower maintenance costs, reduced condensation works, fewer hard-to-let properties, and fewer complaints from tenants, all of which save on staff time as well as direct contractor costs. An increase in asset value may also take place, particularly when the package includes whole house heating, though it may be impossible in practice for the landlord to realise this increase. Some of these estimates are presented in table 4.1. Still wider benefits can be identified, for example to the National Health Service from reduced ill-health in improved housing, but we are not aware of any research which attempts to quantify such benefits.

Table 4.1
Indirect benefits of energy efficiency

Landlord (EEO Case Study No.)	Fuel savings to tenant £/year	Landlord benefit (exc. rent) £/year	Landlord benefit (inc. rent) £/year
Merseyside Improved Homes (155)	105	82	153
Waltham Forest (186)	100	350	not stated
Glasgow (187)	950	300	not stated
Bristol, Wessex House (189)	472	not stated	780
Bristol, Barrowhill Crescent (189)	258	not stated	488

Source: BRECSU 1993a - 1993d

45

Accounting for these indirect benefits is difficult, even where they arise within the landlord's organisation. Costs and savings frequently accrue to different accounts, and are hidden among much larger financial flows. Estimates of indirect landlord benefits nevertheless exceed estimated tenant savings in a number of the cases presented here.

If further research confirms these early estimates in the public sector, it is likely that energy efficiency improvements would proceed more quickly. Government policy on accounting and public spending arrangements will inevitably influence this picture. The situation in the private rented sector is much more difficult. As noted above, the private rented sector contains the largest proportion of very inefficient dwellings. Clearly the identification of landlord benefits similar to those in the public sector could be beneficial, but no literature attempting to develop this area has been identified. In fact the structure of the private sector and the low marketability of energy efficiency in general is likely to work against the emergence of a clear set of benefits to the private landlord. Boardman argues, with respect to all landlords, that a legislative solution is perhaps the most appropriate:

> Thus, with the present legal and economic framework, there is no way of encouraging landlords or tenants to invest in energy efficiency improvements. [ ] The proposal is, therefore, that landlords are made legally responsible for the energy efficiency of their property and any necessary capital expenditure. [ ] Financial assistance can be made available for low-income landlords in the form of a means-tested grant or equity-sharing loan. (Boardman 1991a)

Notwithstanding such a pessimistic view, there is scope for more research in this area, particularly if ways can be found to share the energy saving benefits on an equitable basis.

Energy efficiency and government policy

The political options for improving energy efficiency can be categorised under the following headings:

- subsidies to end users and intermediaries to undertake energy efficiency investment;

- regulations restricting or prohibiting certain energy consuming activities or products;

- measures to increase the price of energy;

- actions to improve the operation of markets in energy efficiency, by providing more, or more believable information;

- actions to remove barriers to energy efficiency by establishing new institutions or by endowing existing institutions with new powers.

The purpose of this section is to examine present policies which aim to encourage energy efficiency, and to discuss possible future policy directions.

Non-public sector policies

The main provision for energy efficiency works by householders at present is the Home Energy Efficiency Scheme (HEES). This scheme provides grant aid to low-income households and covers loft, tank and pipe insulation, draughtproofing to doors and windows and energy advice. The scheme commenced in January 1991 and replaced the Homes Insulation Scheme and the Energy Grant. In its first two years basic measures have been applied to 440,000 households (NEA 1993) and the budget has risen from £26 million in 1991/92 to £37.5 million in 1993/94. Energy efficiency works can also be funded through renovation grants (House renovation Grant and Minor Works Assistance) which are available to people on low incomes. The grants are however discretionary and the amount of money available is dependant on local authority funding limits through the Housing Investment Programme (HIP).

The first 15 months of operation of HEES was reviewed by Neighbourhood Energy Action (NEA 1992) and subsequently updated to cover the first two years (NEA 1993). The review made 24 recommendations and highlighted a number of areas for improvement ranging from the scope of the scheme to administration arrangements. Concern was expressed over the range of works allowed under the scheme (particularly the lack of wall insulation), the allocation of funds, technical quality and supervision, take up of energy advice and levels of customer service. In addition the review identified the difficulties of monitoring progress towards alleviating fuel poverty and pointed to a lack of data on comfort levels and actual energy savings in houses where works had taken place. A study which considered this issue has been carried out by the Building Research Establishment (Oseland and Ward 1993). It was found that actual savings in winter fuel bills averaged £9 compared with a BREDEM estimate, 'fine tuned' to the homes surveyed, of £39. The conclusion drawn from this is that respondents appear to have taken 77 per cent in improved warmth and only 23 per cent in savings. The study is a limited one and this is acknowledged by the authors. The main difficulty is the lack of data on internal and external temperatures. It is quite possible that internal temperatures were low both before and after the works. Actual fuel expenditure suggests that this is likely to be the case since actual 'before' expenditure was 41 per cent less than the expected estimate (which assumed comfortable conditions were maintained for a significant portion of the day) and the 'after' figure 38 per cent less. Hutton et al. (1985) conducted an earlier assessment of the effects of loft insulation and draughtproofing measures carried out by local insulation projects. This too was inconclusive due to lack of data on house and external temperatures and because it was not a before-and-after study (a control group was used for comparison). The present authors are surprised by the absence of well constructed and funded studies in this area.

The Energy Saving Trust and public information campaigns

Other government initiatives outside the public sector have included the establishment of the Energy Saving Trust and the environmental awareness campaign 'Helping the Earth Begins at Home'. The budget for the latter is £15 million over three years and as yet no evaluation of its effectiveness has been published. The goal of the Energy Saving Trust, as set out in its mission statement, is:

> [to promote] the efficient use of all forms of energy in the UK, leading to an overall reduction in total energy consumption and its consequent environmental impact. We will seek to sponsor energy saving initiatives that also provide long-term economic benefits to the nation. (Energy Saving Trust, undated)

The Trust (set up late in 1992) was to be funded through a levy built into the price control arrangements for the privatised energy utilities and through self generated funds for specific projects. Funding in 1994 was around £6 million and was targeted towards schemes involving gas condensing boilers and small scale Combined Heat and Power. In the view of the authors, this level of funding is quite inadequate to the task originally set for the Energy Saving Trust, of making a major contribution to the goal of capping UK carbon dioxide emissions.

Since 1993/94 local authorities have been expected to demonstrate energy efficiency policies which are integrated with the remainder of their housing activity. Assessment of HIP bids will also include energy efficiency criteria. This follows a demonstration programme (the Green House Programme) which provided £60 million over three years from 1991/93 to 1993/94. Overall some 183 schemes were completed covering some 50,000 dwellings and resulted in estimated savings worth £10 million. Annual savings in CARBON DIOXIDE are estimated to be about 50 per cent (DoE 1994b). Details of schemes and general lessons are given in a number of Department of the Environment publications (DoE 1993b and 1994a). The DoE have also published a series of papers (Green House Profiles) which set out the monitoring results of a number of schemes (see for example DoE 1994c).

As a result of the Green House programme and work by BRECSU, detailed guidance is now available to social housing landlords on ways in which energy efficiency can be built into housing policies. The guidance is focused on developing policies which identify the overall energy efficiency of the stock, setting efficiency targets based on affordable warmth, determine the cost effectiveness of the measures adopted and involving tenants as well as professionals in the process. Authorities are encouraged to make use of energy labelling in setting targets and to integrate efficiency measures with modernisation, planned maintenance and reactive maintenance plans (DoE 1994d). Work by the NEA (NEA 1994b) has reiterated much of the advice given by the DoE and has highlighted the value of audits not only in defining cost effective improvements but also in identifying the wide range of efficiencies which exist in what appear to be similar properties. The extent to which authorities overall have responded to the challenge, is still to be assessed although there are useful examples from the Green House schemes (see Green House Profiles) and the NEA study on which replication can be based.

Energy utilities

The use of energy utilities as catalysts and funders of energy efficiency projects has been a feature of the scene in North America for a number of years. Berkowitz et al. (1994) identifies 25 community based projects involving assistance from energy utilities who see conservation as an important part of demand-side management of energy and less expensive than building new power stations. Similar approaches are beginning to be explored in the UK. A scheme of efficiency improvements (including the issue of low energy light bulbs) and advice aimed at managing electricity demand has been carried out in Grantham in Linconshire (Green 1992 and 1994). The role of the energy utilities in promoting and developing energy efficiency was established at the time of privatisation and is enforced by the utility regulators. The further development of their role has been advocated by the Royal Institution of Chartered Surveyors who recommend that the utilities or government provide grant aid which along with administration costs would be recoverable through energy bills (RICS 1994). Possibilities clearly exist, but to date, the Energy Saving Trust is the only broad national initiative identified and that is still in its infancy. Further exploration of the role which could be played by the energy utilities is required.

Home Energy Conservation Act 1995

The Home Energy Conservation Act is set to become the main strategic vehicle for government policy with respect to housing energy efficiency. The Act is expected to come into force in 1996 and will set up energy conservation authorities across the UK. These authorities are to be based on existing housing authorities and it is expected that the existing funding arrangements through the Housing Investment Programme (HIP) will be used to implement energy efficiency strategies. The main thrust of the act is a duty placed on the new authorities to prepare a report which sets out:

> energy efficiency measures that the authority considers practicable, cost-effective and likely to result in significant improvement in the energy efficiency of residential accommodation in its area (Home Energy Conservation Act 1995 s2(2)).

The Secretary of State will have the power to set a timetable for the production of both the initial report and subsequent progress reports. The Secretary of State is also required to provide whatever assistance he considers desirable in the implementation of proposals and may give advice on proposals. In particular, guidance 'as to what improvements in energy efficiency are to be regarded as significant' may also be given. Ministerial statements during the passage of the bill suggest that 'significant improvements' are likely to mean an initial target for energy savings of at least 30 per cent (Seaborn 1995). The act does not specify any particular measures nor does it confer any powers to carry out works or to make grants or loans. However the clear intention is that local authorities should take responsibility for ensuring that efficiency improvements are made in all housing sectors within their area.

The implications of this act, for local authorities, could be extensive in that it will require an assessment of the energy characteristics of the housing stock in order to set targets and establish priorities. It will also require the updating of energy information so that intervention strategies can be monitored and the effectiveness of particular actions determined. As we show with respect to the public sector later in this chapter and illustrate through the case studies in chapters 8 and 9, there are plenty of demonstration schemes which point out the physical changes which are needed to improve both the existing stock and new house designs. For many energy conservation authorities the difficulties will lie, not in deciding on what sort of works are needed but in ensuring that the maximum efficiency gains are made with the resources (both private and public) which are likely to be available.

Local Agenda 21

Local Agenda 21 stems from the 1992 UN Conference on Environment & Development. The objective of the Local Agenda 21 initiative is to push responsibility for meeting the objectives of Agenda 21 down to a local level in signatory states. While one can point to some concrete achievements, the main problem with this initiative in the UK is the narrow, and narrowing focus of local authority activity, which stems ultimately from the absence of a constitutional basis for local authorities themselves. Local authorities are responsible for some housing, education, fire and police services within their geographical areas, and for little else. They have no history of responsibility for the development of broad environmental and energy strategies and no power to implement them. Against this constitutional background, any attempt on the part of central government to devolve the responsibility for identifying and undertaking action on climate change to local authorities is more likely to result in local cynicism than productive activity.

Options for the future

The list of policy initiatives discussed above covers most of the areas listed at the beginning of this section. The main policy category which is yet to be tried in the UK is the deliberate use of the price mechanism, through energy or carbon taxation to encourage energy efficiency (we exclude for the moment the question of duty on transport fuels). The history of the current government's attempts to introduce VAT on domestic energy suggests strongly that carbon or energy taxation is unlikely to be a political possibility in the UK for at least one parliamentary term, and possibly longer. Be this as it may, the authors are of the view that long term energy and carbon dioxide reductions on the scale suggested in chapter 2, are unlikely to be realised against a background of declining real prices for energy.

The arguments in favour of the price mechanism as a tool to increase energy efficiency are that it:

> operates at all levels of the economy, engages all mechanisms by which emissions can be reduced (shifts in mix of consumption, fuel substitution including moves to renewable energy, and technical innovation in all sectors), and is easily collectable. (Lowe 1994).

Unlike most other forms of policy intervention, energy taxation in principle minimises bureaucracy. The main disadvantages of this approach arise if it is introduced suddenly, without a parallel process of education and consensus building, and if it is thought to be a temporary rather than a permanent measure. Under these circumstances, energy taxation is more likely to excite short term responses - transfer of expenditure for those who can afford it, and increased hardship for those who cannot - than the desired response of a general increase in energy efficiency. Energy taxation suffers from the ethical objection that it is regressive, hitting the poor, and occupants of the most inefficient housing the hardest. It is in principle possible to overcome this ethical objection by recycling some or all of the proceeds of such a tax in enhanced social benefits, or reduced taxation, or energy efficiency programmes for those groups of people most affected. A suitable combination of these options could convert a regressive tax into a progressive tax, while increasing still further the overall impact on energy use of such a measure.

It is instructive to compare the UK position on energy taxation with that of the Danish Government. The latter has stated that:

> Energy taxes on both the consumption and supply sides are vitally important for the development of the energy sector. To achieve the desired effect the taxes are in several fields coupled with other means; they are, however, usually the basis for the other means because of their importance for the profitability of investment in conservation and supply.
>
> The Government finds that as a tool of energy policy taxes should remain essential to the attainment of energy targets. Through the use of economic measures based on market forces, it will generally be possible to achieve a more flexible and non-centrally managed development than through the use of administrative measures alone. (Danish Ministry of Energy 1990)

There may be a paradox here, in that market solutions to the problems of energy efficiency may only be politically feasible in societies with a high degree of social cohesiveness, income equality, and a relatively secure and generous provision for the long term sick, the out-of-work, and other groups who, in the UK, are most prone to fuel poverty. Governments which generally pursue an aggressive free market philosophy, may find that market solutions are unavailable to them in the area of energy policy because of the accompanying social conflicts.

Conclusions

Any attempt to devise policy to promote energy efficiency in housing must confront the significant differences that exist between different parts of the sector. At its starkest, for the rich, energy efficiency is about using less energy and saving money. For the poor, it is about using energy to better effect, raising comfort standards, and avoiding ill-health from poor housing conditions. These issues are discussed further in chapter 6. Standards of energy efficiency will need to rise considerably in the poorest sections of the population, before further improvements in efficiency lead to a significant fall in energy demand. Most of the short term savings potential that exists in this section of the population arises from fuel switching rather than from reductions in delivered energy (for an example the reader is referred to chapter 8). Policies based on market mechanisms (provision of information, rating schemes and pricing policies) are likely to work in the former section of the population, but not the latter.

Effective policy making requires coherent rafts of policies, and careful and detailed implementation based on a respect for the personal interests and the social, political and institutional mechanisms that operate in each part of the housing sector. It must also be based on an understanding of human behaviour, and of the underlying cognitive processes. For a discussion of these matters, the reader is referred to chapter 7.

5 Land-use planning and energy efficiency

Introduction

Sustainable development is a matter of great significance for many aspects of land-use planning. This is even more the case in urban areas where, as the World Commission on Environment and Development (1987) notes, 'by the turn of the century, almost half the world will live'. Contained within the concept of sustainable development are a number of complex and interlocking arguments. Central to this coincidence of features and factors is the role of cities as places of settlement and habitation. Housing is at the heart of the phenomenon of urbanisation, and it represents touching on all three fundamental reasons for the very existence of settlements; security, shelter and social exchange.

Cities as centres of population and housing are major consumers of resources. Water, food, energy and other raw materials are consumed in large quantities by urban activities. Whilst part of this consumption is required in order to satisfy the needs of industry and to allow for the operation of the services that are provided by cities to other places, a substantial element of the resources consumed reflect the needs of the urban system itself. Even though some cities may consider that they do not have a major direct impact upon the environment (for example because activities such as electricity generation or coal mining are not carried out within the boundary of the city), they nevertheless cast a significant environmental footprint or ecological shadow (MacNeill, Winsemius and Yakushiji 1991) over other parts of the earth's surface. The modern city is a significant exporter of environmental impact. The massive use of energy by cities is a major element in this uneven balance between the production and the consumption of resources.

Taking one major metropolitan area as an example, Greater Manchester accounts for an estimated 1/700th of world demand for fossil energy (Ravetz et al. 1995), and some 30% of this energy is used by the domestic sector (see table 5.1). A further 22 % of energy is consumed by transport, with the balance of 48% used by commerce and industry.

Table 5.1
Energy use totals in Greater Manchester

Delivered Energy to final users, 1000's GWh

Sector	Oil	Coal	Gas	Electricity	Total	Percent
Domestic	5	40	176	45	266	30
Transport	193	0	0	2	195	22
Commerce	51	5	44	33	133	15
Industry	64	47	120	61	293	33
Total	313	92	340	142	887	100

Source: Ravetz et al. 1995

Given the high percentage of energy consumed by housing, strategies for energy reduction in cities, have to address three key questions:

- Can housing and other buildings be made more energy efficient?

- Can reductions be achieved in the movement of goods and materials and can other transport movements be made more energy efficient?

- Can fewer environmentally damaging forms of energy production be substituted for current methods of production and use?

Some of these questions are addressed in other chapters. Answers to others would require an analysis which is beyond the scope of this book. However, it is important to acknowledge that a solution to the urban energy question cannot be achieved through piecemeal action, rather it has to be arrived at through the design and implementation of a comprehensive and integrated approach to the overall functioning of the urban system.

Although the focus of much of the recent research on energy and land-use planning reflects the problems encountered in urban areas, a range of other non-urban issues can be identified. Rural areas demonstrate many of the characteristics of resource consumption that are attributed to cities. Indeed some have developed settlement patterns that are dispersed, with high levels of personal movement and energy use, and resemble urban areas such as Los Angeles more than traditional rural areas. Even though the majority of policy initiatives on sustainable development in general, and on energy efficiency in particular, concentrate attention on the functioning of urban systems, a growing understanding and awareness has emerged that implies the need for the development and application of a comprehensive approach to policy that encompasses both urban and rural areas. Planning the sustainable town or region requires intervention at each scale in the urban hierarchy from the individual dwelling to the metropolis (Breheny and Rookwood 1993) and implies a form of intervention that is 'devised and implemented in an integrated, complementary fashion'.

This new awareness that policy has to be constructed and implemented in an integrated manner - from global to local - is reflected in a series of policy documents and statements at national and international levels. Agenda 21 represents the consensus reached by 179 nation states at the United Nations Conference on Environment and Development, held in Rio de Janeiro in 1992, and amongst the agreements, energy matters were considered to be of great importance. The sound management of human settlement was seen to require actions to promote energy-efficient housing design and more efficient transport (Keating 1993). Similar messages can be seen in the European Union's programme 'Towards Sustainability' (Commission of the European Communities 1992), and in the UK Strategy for Sustainable Development (UK Government 1994). But in order to ensure that implementation follows strategy, policy making and elaboration must cascade down the spatial hierarchy to regional and local levels.

Translating global concerns into local action goes further than the traditional aspirations of conventional town and country planning. It also implies taking action at the level of the individual household. Amongst such actions are waste minimisation, the conservation of resources and enhanced energy efficiency. The remainder of this chapter concentrates attention upon this final issue.

Planning, energy and housing

Within the greater awareness that planning has an important role to play in the promotion of energy efficiency in the design and operation of human settlement, it is possible to identify a number of broad organising concepts and practices. The most important of these are:

- the need to develop methods of integrated analysis and plan-making;

- the desirability of treating the urban system as a whole, and inter-relating the major factors that influence the location and operation of settlements;

- the need to introduce methods for the strategic assessment of the potential impact of policies, programmes and plans.

The first of these topics, developing methods of integrated analysis and plan-making, has been investigated by a number of authors. Healy and Shaw (1993) identify five features of a regulatory land-use planning system which recognise amongst other environmental factors, the energy issue. Such a system requires mechanisms for:

- identifying the full range of local site-related matters that are relevant to fostering locational patterns which minimise energy use;

- ensuring that development does not exceed ecological carrying capacity thresholds;

- balancing environmental, social and economic considerations in development;

- promoting and managing local environments and for;

- dealing specifically with the adverse impacts of development.

The third of these points, balancing development, is reflected in other research (Roberts 1995) and is considered later in this chapter.

An essential requirement for the promotion and achievement of greater energy efficiency in human settlement is the need to ensure the effective integration of all relevant sectors of activity. For example, whilst the promotion of more energy-efficient housing is clearly desirable in itself, it is also essential to ensure that the inhabitants of such housing have access to an effective system of public transport. Local co-ordination and joint responsibility are vital elements in ensuring that the public, private and voluntary sectors work together, and that departmental boundaries are permeable in order to allow for the best solution overall (OECD 1991).

The third feature noted above relates to the need to prevent rather than cure environmental problems. Environmental impact assessment (EIA) is a procedure that aims to identify and eliminate, often through negotiation, the negative environmental consequences of proposed developments. The majority of applications of EIA have been at the level of an individual site and relate to a single project. Whilst this is helpful in eliminating the adverse consequences associated with a particular project, EIA is more likely to be effective if it is applied at higher levels in the policy hierarchy. Strategic environmental assessment (SEA) seeks to avoid the emergence of policies, programmes and plans that may cause environmental harm, and to minimise the wasteful use of resources that is associated with the generation of detailed project proposals that may be rejected at the final hurdle due to their inherent environmental inefficiency. Glasson (1995) notes that good practice is now spreading following the promotion of SEA by the UK Department of the Environment and its adoption by local planning authorities.

A system of SEA designed to promote greater energy efficiency in housing would, for example:

- allow the development of national policies that specify housing types, patterns of development and suitable infrastructure;

- encourage the integrated development of public transport systems at all scales, especially in urban areas;

- support planning proposals that allow for greater consideration to be given to a range of energy-efficiency criteria in allocating land for housing;

- promote the generation of a greater awareness among developers of the options available to them in the design of specific houses and the configuration of housing developments.

Although some of these ideas and operational practices are relatively recent, there are a number of lessons and messages from established sources that illustrate the actual or potential influence of land-use planning over energy efficiency in housing. Despite the fact that many of the earlier reviews of theory and practice have tended to emphasis broader, more strategic aspects of the interface between energy and land-use (Watt Committee 1979), or have examined the general characteristics and treatment of energy efficiency in the planning and management of settlements (Elkin et al. 1991), a number of other issues have been examined in research projects.

As has been noted above, it is possible to identify an important and growing literature that is more directly concerned with the relationship between land-use planning and energy efficiency in housing. This literature can be considered in three broad categories:

- issues related to situation, site, aspect and orientation;

- factors related to the development pattern of neighbourhoods within urban areas, the establishment of efficient energy infrastructure, and the encouragement of local communities to engage in energy efficiency; and

- broad questions associated with the overall management of urban systems and the spatial distribution of the full range of urban functions.

Each of these categories, together with a number of specific topics, are now considered in more detail in the following sections.

Situation, site, aspect and orientation

The first issue to be considered relates to the situation of housing. This factor reflects the desirability for housing to be situated so as to achieve an optimum or near-optimum relationship with other urban activities. This relationship may be expressed in terms of the geographic features of a situation - a linear feature such as a road may influence the location of housing - or the functions associated with a particular location, and the provision of retailing or leisure facilities. Linear features have exerted, and continue to exert, considerable influence on housing distribution. Although some authors have concluded that such factors have energy advantages (Owens 1986), they also state that they are not as well defined as the benefits of greater compactness or functional integration. Furthermore, there is an important need to balance the energy efficiency of particular linear or compact features with questions of social welfare (Breheny 1993). There is little point in promoting greater energy efficiency through the cramming of cities, if the result is an increase in social stress or negative socio-economic responses.

One must recognise that while high density may minimise environmental impacts that arise from certain categories of transport, other impacts may increase with density. There is, for example, some evidence that recreational transport demand increases at high density. Situation is also important in relation to the question of trip minimisation. Significant dispersal of functions may tend towards greater trip generation. This favours the creation of integrated rather than fragmented communities, although fragmentation may not be a disadvantage so long as the fragments are joined by an effective communications network (Goodchild 1994). High density may also militate against strategies for minimising environmental impact from water supply, sewage treatment and disposal of bio-degradable wastes from dwellings. The pioneering Southwell House (Bunn 1994) shows that it is technically possible for a house to supply all its own water, and to compost its own sewage and food wastes on a semi-urban site of approximately 1/20 hectare. It is not clear that this would be possible at densities much higher than this.

The arguments presented here suggest that the influence of density on environmental impact is not at all simple. Factors such as spatial integration of functions, the number of possible destinations within easy travelling distance and the presence of high-speed road or other transport networks, are important modifying variables. It appears that environmental impact may be a 'U' shape, rather than a monotonically decreasing function of density, with a wide and, as yet, poorly defined minimum.

Site is a fundamental issue, especially in relation to the functions discharged by a particular urban form or area of housing. The choice of a site for housing is likely to be influenced by historic factors that are reflected in the development (these factors may themselves be energy efficient), or by new factors that attach particular important to energy efficiency. It is clearly impossible and would be energy inefficient, even it were possible, to bring about wholesale alterations to an existing pattern of settlement. Extended or new settlements have to be designed within the context of the established network of places (Locke 1994), but in doing this it is vital to identify new housing sites that represent the maximum degree of energy efficiency. This approach has been reflected in recent revisions to statutory plans (Bedfordshire County Council 1994) and is advocated by the Department of the Environment (DoE 1993c).

There are many further potential gains obtainable through the adoption of more sophisticated policies which either seek to minimise the consumption of environmental resources, or ensure that policies do not conflict with the needs of the environment through the advance application of environmental assessment (Glasson 1995). The adoption of strategic environmental assessment can help to create a framework of policy within which criteria can be defined that can be used to guide local planners in seeking the most energy efficient overall use of land and of a particular site.

Aspect is a traditional concern in land-use planning and reflects the long-established practice of attempting to maximise the use of favourable slope patterns. Traditionally this has to do with maximising the availability of daylight, although in recent years greater emphasis has been placed on the potential for solar gain (Vale and Vale 1993). Although the potential for solar gain is limited in the UK, it is of greater significance in other countries, where local and regional planning is seen to have an important role to play in ensuring that the allocation of land and the deployment of activities at a site level makes best use of any solar potential. Although less significant in the UK, passive solar gain is, nevertheless, still an important element in the overall picture of energy efficiency. Given the choice of enhanced aspect or not, it would be foolish to ignore the potential for passive solar gain (Breheny and Rookwood 1993).

The orientation of housing on a site that has a favourable aspect is also crucial. Passive solar energy can provide up to 20 per cent of the annual space heating energy required for a well insulated building (Vale and Vale 1993), although this may conflict with other desires and objectives, such as the need for privacy or security (Lowe et al. 1985). Site, aspect and orientation factors can be combined with novel construction techniques, such as earth-sheltered housing design, in order to further enhance energy efficiency (Blunden and Reddish 1991), although innovations of this nature may require careful and patient negotiation with development control officers and building inspectors.

A useful summary of site, aspect and orientation factors is provided by Table 5.2 (from Keplinger 1978). This indicates the wide range of possible factors and considerations that should be taken into account when deciding upon the location and layout of housing in order to maximise energy efficiency.

Above and beyond the specific consideration of situation, site, aspect and orientation, it is also important for planning to take account of a range of other matters that relate to energy efficiency. Such factors as the control of wind speeds by constructing wind breaks close to walls, or using buildings as wind breaks themselves, can be important in enhancing energy efficiency. The choice of appropriate housing types, design and layouts are all important factors in influencing local microclimate (Rydin 1992).

Indeed, the way in which a development interacts with its microclimate will have important implications for energy efficiency. The design of any development will modify the microclimate around it and the extent of this modification is, at least partly, in the hands of the designer. An extremely useful review of design considerations as they relate to the UK has been provided by the Building Research Establishment (BRE 1990). The main development principles and considerations put forward in this document are summarised in Table 5.3.

Table 5.2
Site orientation chart

	Cool regions	Temperate regions	Hot humid regions	Hot arid regions
Objectives	Maximise warming effects of solar radiation, reduce impact of winter wind, avoid local climatic cold pockets	Maximise warming effects of sun in winter. Maximise shade in summer. Reduce impact of winter wind but allow air circulation in summer.	Maximise shade. Maximise wind.	Maximise shade late morning and all afternoon. Maximise humidity. Maximise air movement in summer.
Adaptations				
Position on slope	Low for wind shelter	Middle-upper for solar radiation exposure	High for wind	Low for cool air flow
Orientation on slope	South to southeast	South to southeast	South	East southeast for afternoon shade
Relation to water	Near large body of water	Close to water, but avoid coastal fog	Near any water	On lee side of water
Preferred winds	Sheltered from north and west	Avoid continental cold winds	Sheltered from north	Exposed to prevailing winds
Clustering	Around sun pockets	Around a common sunny terrace	Open to wind	Along E-W axis, for shade and wind
Building orientation	Southeast	South to southeast	South towards prevailing wind	South
Tree forms	Deciduous trees near building. Evergreens for windbreaks	Deciduous trees nearby on west. No evergreens near on south	High canopy trees. Use deciduous trees near building	Trees overhanging roof if possible
Road orientation	Crosswise to winter wind	Crosswise to winter wind	Broad channel, east-west axis	Narrow, east-west axis
Materials colouration	Medium to dark	Medium	Light, especially for roof	Light on exposed surfaces, dark to avoid reflection

Source: Keplinger 1978

Table 5.3
Considerations of microclimate in site development

Main principles

Practical considerations

Solar access:

In addition to warming internal spaces, solar radiation can also have a warming effect on the spaces around buildings which contributes to the 'heat island effect' which is observed in built up areas. The more of this energy which can be retained during the heating season, the less energy will be needed to heat individual dwellings. Summer overheating may be a problem if adequate shading is not provided.

- Arrange road layouts and plots to enable wide south facing frontages to enhance solar access to individual houses and the spaces between them.
- Avoid/minimise overshading by buildings and other site features. e.g. leave southern ends of courtyard groups open, arrange 2/3 storey houses to the north of single storey houses.
- Materials with good radiation absorbtion properties and high thermal mass will enhance the effects of solar radiation.
- Shading devices on individual houses and groups can reduce summer overheating, also deciduous trees can provide summer shade with modest winter solar access.

Wind speed:

Reducing wind speed around buildings ensures that direct heat loss is reduced and that higher air temperatures in the spaces between them are not rapidly displaced by colder air. Reducing wind speed can also improve outdoor thermal comfort. Sheltered sunny spaces can be comfortable even in winter. This will require attention to the shape and arrangement of building groups as well as the designing of specific shelter features.

- An assessment of the magnitude, direction, temperature and duration of winds will be required for a site.
- Avoid 'wind channels' formed by long parallel terraces. Long straight transport routes can create wind channels resulting in significant discomfort for pedestrians as well as increasing heat loss from adjacent houses.
- Arrange building groups to maximise shelter.
- Design shelter structures to give shelter to the whole development and local areas. If the site is naturally sheltered, make the most of it.

Driving rain:

Building surfaces which are wet for long periods will enhance heat loss through evaporation. Wetting of surfaces is a function of wind speed as well as rain fall. These two factors can be combined into an index of driving rain. The extent of this effect will also depend on the porosity of surfaces. Low surface porosity will in less evaporative heat loss.

- Many of the considerations in providing wind shelter are applicable here, but driving rain should be considered as a separate issue. This is because there will often be differences between the direction of rain bearing winds and wind directions important for other aspects of shelter design. It may be necessary to strike a balance between conflicting directions.

Source: BRE 1990

It is rarely possible to achieve a perfect combination of site, aspect and orientation, due to other planning constraints. In a highly populated country, it is not possible only to build on sites with an ideal aspect. Solar access on a given site becomes harder to achieve at very high densities, and the need to provide developments with a sense of enclosure and intimacy places some constraints on pure passive solar layouts. The most important influences on space heating energy use in dwellings are, in any case, to do with the insulation and air-tightness of the thermal envelope of the individual dwelling, and given this fact, the single-minded pursuit of passive solar gain would be mistaken. The best of the passive solar projects of the 1980s nevertheless show that modest solar contributions are entirely consistent with other planning objectives, and that site planning has a part to play in the search for energy efficiency, particularly in new dwellings. The keys are an understanding of the technical issues, and a willingness to seize opportunities for passive solar design when they present themselves.

Development, infrastructure and empowerment

The recent growth of concern amongst planners for the promotion of greater energy efficiency is linked to the search for methods of planning that can assist in the promotion of environmental sustainability. Energy is at the centre of many investigations of sustainable development (Blowers 1993 and Breheny 1992) and, above and beyond the issues considered in the previous section of this chapter, a number of other important themes can be identified.

The first of these themes, is the search for methods of operating that will ensure greater integration between land-use planning and the planning of energy supply. In some countries the process of integration is deeply-rooted, especially in those countries that have few indigenous sources of (non-renewable) energy (Blunden and Reddish 1991) or where energy is, and has always been, expensive. Since the 1920s many communities in Denmark have used the 'waste' heat from diesel engine power plant in combined heat and power or district heating schemes. This has led to the insertion of energy priorities into planning procedures and, in some cases, the establishment of land-use plans that are directed towards the achievement of energy efficiency objectives (Christensen and Jensen-Butler 1980). Neighbourhood layouts may, for example, be dictated by the technical requirements of combined heat and power and district heating, creating patterns of urbanisation that allow for the economic and technical optimisation of the district heating system.

A second theme reflects the increased emphasis placed upon the desirability of integrating the residential, commercial, industrial and other functions of an urban area. This implies that gains can be achieved through the integration of sectoral strategies for energy efficiency and, for example, residential energy requirements can be met from waste heat produced by industry. The interaction between business and a local community which is implied by this transfer of what would otherwise be wasted heat, requires the adoption of a partnership approach to planning (Roberts 1995). A partnership approach may also help in the design and implementation of a range of other environmental initiatives.

Planning at the neighbourhood or district level may also be influenced by other energy-related considerations, including the establishment of waste incineration facilities linked to the generation and supply of energy, and the assignment of a high degree of priority to the layout of housing in order to allow the maximum use to be made of public transport. The overall aim of an integrated system of land-use and energy planning is to enable the best combination of factors to be identified, rather than to optimise on a single variable to the exclusion of others (Owens 1992). The most suitable combination should match the characteristics of the individual neighbourhood, city block or urban area and it is likely that variety rather than uniformity will emerge.

Integration should be based upon an array of energy objectives that reflect the need for the urban system as a whole, similar to the 'balanced development' objectives of sustainable development (Roberts 1994). This reflects the view that 'global warming, like the U-boat blockade, makes for awareness that resources are limited and should be husbanded' (Hebbert 1992). Housing interacts with the other elements of an individual system in many ways, and it is difficult, if not impossible, to identify a single model that can guarantee success in all locations (Breheny 1992).

Infrastructure provision for greater energy efficiency in housing is equally diverse. The provision of the infrastructure necessary for a combined heat and power and or a district heating scheme has been discussed above and these requirements have been reflected in the literature concerned with the development and implementation of such schemes (Hutchinson 1990 and Environment City 1993). Urban transportation improvements are equally important, both as a way of increasing the energy efficiency of household movements by public transport (Commission of the European Communities 1990), and as a means of minimising the need for the use of private vehicles. A difficulty encountered in many areas is that of obtaining the finance necessary to fund improvements in infrastructure which will lead to greater energy efficiency; this problem is faced by many local authorities and can be seen to have delayed the implementation or caused the abandonment of suggested improvements (AMA 1985).

A key factor which determines energy efficiency in many urban and rural areas is the distribution, availability and accessibility of transport. These are matters now increasingly taken into account in the preparation of plans and the results of innovative policy can be dramatic. If, for example, accessibility to public transport can be improved by ensuring that frequent and reliable public transport is available at no more than four to five hundred metres from any house (assumed to be the maximum distance that most people are willing to walk), then a substantial shift from private to public transport may occur. Currently, the private car accounts for some 71 per cent of journeys, with public transport accounting for only 17 per cent (see Table 5.4).

If car users can be persuaded to either shift to the more efficient forms of public transport or, at worst, to switch to a more energy-efficient car used at a higher load factor than average, then substantial energy gains can be expected. In order to promote the greater use of public transport, it may be necessary to adopt new methods of planning. In searching for new areas of land for housing, for example, it would be sensible to define the area of search by reference to the presence of existing rail or bus routes.

Table 5.4
Percentage of travel by transport mode

Mode	Passenger Mileage	Time Spent on Journeys
Walk	5	28
Bicycle	1	1
Motorcycle	1	1
Car Driver	)	)
Car Passenger	) 71	) 48
Van/Lorry	4	)
Local Bus	6	12
Private Hire/Tour Bus	3	2
Train	6	5
Coach/Express Bus	1	)
Other Public Transport	1	) 2
Other Private Transport	1	1
Total	100	100

Source: Potter and Hughes 1990

As demonstrated in a recent report from the Royal Commission on Environmental Pollution (1994), there are significant differences in energy consumption between transport modes. For all transport modes, load factor is an important determinant of energy consumption (see Table 5.5).

It is important to recognise that there are significant interactions between transport modes. Motorised road transport in particular tends to inhibit the most energy efficient forms of personal transport, cycling and walking. The widespread ownership of the private car creates a powerful dynamic at all levels from the personal to the regional, which encourages the spatial separation of functions, and reduces the number of possible destinations within walking or cycling distance. The personal danger to pedestrians and cyclists from powered vehicles is also a major factor in supressing walking and cycling, particularly among the young and elderly. The effect of these considerations is that simple comparison of direct energy intensity may well underestimate the overall energy and environmental impact of motorised road transport.

Table 5.5
Energy efficiency of transport by mode
(energy consumption in MJ per passenger km)

Mode	At typical occupancy	Fully Loaded
Express Coach	0.3 (65%)	0.2
125 train (diesel)	0.8 (50%)	0.4
255 train (electric)	1.0 (50%)	0.5
Small diesel car (1.8 ltr)	1.2 (35%)	0.4
Small petrol car (1.1 ltr)	1.4 (35%)	0.5
Suburban train	1.7 (22%)	0.4
Large diesel car (2.5 ltr)	1.8 (35%)	0.6
Large petrol car (2.9 ltr)	2.8 (35%)	1.0
Air, internal flights	3.5 (65%)	2.3

Source: Royal Commission on Environmental Pollution 1994

A final important point relates to the 'ownership' of proposals to enhance the energy efficiency of housing. This suggests that people respond best to calls for greater action on energy efficiency if they are involved in the planning and implementation of solutions, and if energy efficiency is linked to their personal concerns. One author has suggested that 'a green-housing policy which does not recognise poverty isn't worthy of the name' (Brooke 1991). One possible response to this concern is a decentralised approach to energy efficiency. Evidence of the gains that are associated with a decentralised approach can be seen in the work done on community energy planning, especially in North America (Cullingworth and Sparling 1988).

Urban management

Many of the issues related to the promotion of energy efficiency in housing are interlinked with questions connected to the achievement of a more sustainable urban form. This central concern has been examined from a number of standpoints: urban modellers have developed ecological analogies (Xu and Madden 1989), spatial analysts have discussed the merits of compact and dispersed cities (Goodchild 1994), and governments have offered a range of policy prescriptions. A distinguishing feature of the debate, as it affects energy efficiency in housing, is that there is no single optimum solution. As noted above, most evidence suggests that a desirable combination of factors, rather than a single model, is most likely to lead to improvements. The variety of urban types, and the wide differences that occur in terms of microclimate, economic opportunity and housing stock, suggests the need for solutions to be tailored to the characteristics and features of each urban area. Within particular areas, specific solutions will have to be developed which reflect the characteristics of individual houses, groups of houses or high-rise blocks.

However, a number of guiding principles can be identified that can be used to help to create and promote strategies that are aimed at improving the energy efficiency of housing. Some of these principles relate to the management of energy efficiency:

- the establishment of quantity management systems;

- the development of indicators to define the pattern of resource use and maximise progress;

- the creation of consensus - building that allows for agreement on the need for action and implementation (Lusser 1994).

Other principles are concerned with enabling greater energy efficiency through the creation of integrated environmental plans that consider, amongst other factors:

- urban form and energy consumption;

- promotion of public transport use;

- promotion of CHP;

- telecommunications and non-physical interaction;

- recycling schemes;

- building controls;

- urban densities (Breheny and Rookwood 1993).

Whilst such principles do not provide a specific solution, they do offer a framework for the consideration of energy efficiency and, more importantly, they offer the prospect of an overall solution. There is, for example, little point in promoting more energy efficient housing if the residents of such housing still have to travel an excessive distance to work by private transport. Energy efficiency in housing is a major important element of creating a sustainable city, but it cannot be considered in isolation.

An attempt to apply the concept of the sustainable city region to a specific metropolitan area is currently underway. This research aims to build upon the work of the Town and Country Planning Association's Sustainable Development Study Group (Blowers 1993). It is also informed by the SERC/ESRC Cities and Sustainability Study (Elkins and Cooper 1993), and the report of the OECD's Project Group on Environmental Policies for Cities in the 1990s (OECD 1991). The city region selected for the application of the concept is Greater Manchester, and energy is one of the key factors under consideration. Amongst other questions under investigation are:

- Production: How can the impacts of energy production be reduced and what are the cost and land-use implications?

- Local Forces: How far can a city region become self-sufficient in terms of local renewable energy production, after the reduction of demand through conservation?

- Integration: What is the potential in urban centres for the promotion of integrated energy generation and consumption technologies such as CHP?

These questions, together with parallel investigations of other major functional sectors, require answers in order to help to construct a range of scenarios that will be used to create a pathway towards the achievement of sustainable development (Ravetz 1994). It is studies such as this that will help to frame the future land-use planning context for energy efficiency in housing.

Conclusion

This chapter has demonstrated the importance of considering energy efficiency and housing as part of the broader debate on the implementation of land-use planning strategies for sustainable development. By adopting such

69

strategies the question of energy efficiency can be integrated into the design and development of the entire urban system.

All scales of planning are affected by this move to a new and more sustainable model. At international and national levels it implies the generation of policies which acknowledge the need to avoid modes of development which necessitate additional increments of energy use in order to achieve both economic development and social objectives. At regional and local level it implies designing with the environment rather than superimposing a form of urbanisation that cuts across the characteristics of the physical environment.

Above all, such solutions have to be strategic, long term and integrated. Considerable benefits can be achieved by working in partnership to agree and implement development agendas which serve the needs of the entire community and which allow for the generation of both economic and energy efficiency.

6 Energy, poverty and health

Introduction

We have, so far, discussed the need to conserve energy largely in terms of the environment and global, national and local issues. This chapter and the next consider questions which relate to individual houses and their occupants. In particular, it examines the human problems which result from energy inefficient housing. The inefficiency of much of our existing housing stock means that large amounts of energy are required to maintain comfort temperatures. In many cases comfort temperatures are difficult to achieved despite considerable expenditure on fuel. The same is true, to a lesser extent, of domestic hot water systems. For those on low incomes (and even for some on higher incomes), this often leads to considerable hardship as they are unable to afford the energy costs of providing warmth. Many of the more general problems associated with poor housing are related to the energy issue. Houses which people cannot afford to heat often suffer from dampness caused mainly by condensation, which, in addition to being a problem for the occupants, also contributes to increased deterioration of the building fabric.

Cold and damp houses which are prone to disrepair, are not just uncomfortable to live in, they also represent a threat to the health of the occupants. Respiratory, heart and allergy problems have all been associated with such conditions. This is particularly so in the case of vulnerable groups such as the elderly, the very young and the disabled. Even without the environmental problems outlined in chapter 2, the social and health problems associated with poor and inefficient housing would provide a strong case for a vigorous energy efficiency policy in housing. It is to these issues that we now turn.

Energy efficiency and poverty

Although the ability of the poorest in society to afford adequate warmth and shelter have been matters of enduring concern, the oil crisis of the early 1970s brought the notion of fuel poverty into focus (Campbell 1993). Concerns during the 70s and early 80s seemed to concentrate on the problems of fuel debt and disconnections as problems in their own right which should be addressed by social policy. The early studies of fuel debt in Manchester in 1973 (Hesketh 1975) and in Leeds and York in 1975 (Richardson 1978) pointed to the importance of both low income (contrary to some popular notions at the time) and the high level of fuel costs as important factors. In the Manchester study it was found that weekly fuel costs were at the top end of the expenditure range as indicated by the Family Income Expenditure Survey for the same year and that families claiming benefit spent about 40 per cent more on heating than was allowed in the notional fuel element of the benefit. Clearly fuel debt and disconnections are useful indicators of fuel poverty but as Parker (1983) points out:

> While the growing numbers of disconnections are disturbing [ ] it must be emphasised that fuel debts are only one part of a much wider increase in indebtedness.

It must be pointed out also that not everyone in fuel poverty goes into debt or suffers disconnection. The elderly in particular are more likely to skimp on fuel and to endure cold conditions rather than suffer the ignominy of debt and disconnection. Disconnection rates are also lower among the elderly due to the more lenient policies of the fuel utilities towards this group (Berthoud 1983). Both fuel debt and disconnections are important areas in their own right and have wide implications for social policy, disconnection policies of fuel companies and the provision of debt counselling. The relationship with energy efficiency however lies in the realisation that by increasing the energy efficiency of the dwellings of the poorest households it should be possible to enable them to obtain warmth at a price that they can afford.

The idea of tackling the problem from a conservation and efficiency standpoint has been discussed by Bradshaw and Hutton (1983) along with other suggestions on income and tariffs. The most significant work in this area is however that of Brenda Boardman (Boardman 1991a). Boardman's analysis sets out quite clearly the importance of understanding both the income and expenditure side of the problem and the crucial importance of energy efficiency in enabling the two to meet. Most importantly her approach tackles the critical question of providing an adequate and affordable level of

72

the real commodity, warmth. A policy of addressing the state of the capital stock would represent better value for money than providing long term support through the benefit system.

Put simply; people on low incomes have great difficulty in paying fuel bills, are usually cold, and are unable to invest in improvements which would result in higher temperatures or lower bills or both. Boardman's concluding definition of fuel poverty emphasises the energy efficiency aspects:

> Fuel poverty is the inability to afford adequate warmth because of the energy inefficiency of the home. (Boardman 1991a)

Clearly the energy efficiency of a home has a large influence on the ability of someone on a low-income to afford adequate heat. However fuel poverty is also a function of the level of income and it is important to understand how the two interact. The model proposed by Boardman is one of 'Affordable Warmth'. This begs two clear questions; What is affordable? and What constitutes warmth?

Affordability

The starting point here is a definition of poverty. The poor are defined by Boardman (1991a) as those UK households who depend upon the state for at least 75 per cent of their income and the proportion of households in this category are estimated to be about 30 per cent. Moreover it is argued that this group can be modelled with reference to the bottom 30 per cent of the income range as identified in Family Expenditure Survey statistics. Using this proxy group it is concluded that the poor spend (on average) about 10 per cent of their income on fuel (about 50 per cent of which is on heating). In 1989 this amounted to an average of £8.36 per week (Boardman 1993). It is this proportion (10 per cent) which is used as the notional level of what is 'affordable'.

Although 10 per cent acts as a point of reference it should be remembered that this average masks considerable differences between certain groups and individuals all of whom are making a series of personal judgements about affordability and the value placed on warmth. In 1988 single parents allocated 15.8 per cent of income to energy and pensioners 14.4 per cent. This compares with 4.7 per cent for the country as a whole (Hunt and Boardman 1994). Given the state of the existing housing stock (and the poor generally occupy the least efficient houses) and current levels of expenditure, the poor are not able to achieve adequate levels of warmth. To achieve adequate levels by increasing income (social security payments and pensions) could add over

£4 thousand million to annual expenditure and increase annual emissions of carbon by some 13 Mte (Hunt and Boardman 1994).

Warmth

Warmth is a concept which relates primarily to comfort. This therefore requires some understanding of the way in which thermal sensation is perceived by people. There are six main variables used to assess thermal comfort. These consist of four physical environment variables: air temperature, radiant temperature (the extent of radiant heat exchange between a body and surrounding surfaces), air speed and humidity, and two variables related to personal factors, which are the level of activity (determining heat production within the body) and the type of clothing. The same feelings of thermal comfort can be achieved by an almost infinite range of different combinations of these variables. In houses where air speeds are low and humidity is less than the recommended 70 per cent relative humidity (RH), air temperature is often taken as a reasonable measure of the physical environment.[1] This assumes however that mean radiant temperature is roughly the same as air temperature (within 2°C - see Markus and Morris 1980 for a full discussion of the principles of thermal comfort). Comfort is not a matter of physical sensations alone however. The very notion of comfort is a psychological one rather than a physical one. Recent work on comfort (Baker and Standeven 1995) is beginning to recognise the likelihood of a significant contribution to comfort which can be made by ensuring that people have control over, or have the freedom to adapt to, their environment. Although only dimly understood, this may have important implications for the design of houses which aim to provide affordable warmth.

Notwithstanding the complexities of the comfort issue, design recommendations are almost invariably quoted in terms of air temperature targets. For example, recommendations for the design of new housing require the achievement of 21°C in living rooms and 18°C in bedrooms (BS 5449 1990) and these figures correspond with the general levels of between 18°C and 24°C which have been identified by the World Health Organisation (1984) as posing little threat of thermal stress to sedentary individuals with air speeds of less than 0.2 m s^{-1} and a relative humidity of 50 per cent. In temperature controlled environments, there seems to be little difference in feelings of thermal comfort between the old and the young (Collins 1993). However the lower physical activity levels of the old may result in the need for temperatures to be some 2 or 3°C higher. Given the comments discussed above about the importance of users being able to control their thermal environment, the needs of the elderly for warmer temperatures may also relate to the reduced opportunity for control in some environments designed for the elderly.

Defining adequate temperatures for health is, if anything, harder than for comfort. It is generally accepted however that the risk of respiratory problems rises as temperatures fall below 16°C. This figure was enshrined in the Offices Shops and Railway Premises Act 1963 although its precise basis seems to be a little obscure (Collins 1993). Be that as it may, there is clear physiological evidence of changes in blood pressure and blood constitution as temperature declines below 15°C (Collins 1993, Collins et al. 1985, Keatinge et al. 1984). These issues are discussed later in this chapter.

In order to provide a single measure of warmth it is necessary to seek a whole house average temperature which takes into account not only the desired comfort temperatures in each type of room but also the time spent in the house. The use characteristics of low income households have been discussed by Boardman (1991) and when combined with temperature standards the following broad calculation results:

a temperature of 21°C in two rooms, kitchen and bathroom for 13 hours during the day for the first occupant, and an additional room for each person when they are in the home. At night, 16°C in the bedrooms of people vulnerable to respiratory disease. Other unoccupied rooms to be maintained at 14°C to prevent condensation, mould growth and thus, ill health. This gives a 24-hour, whole house MIT [mean internal temperature] of about 18°C. (Boardman 1991a)

In applying her model to an example case, Boardman (1993) calculates that an energy audit rating of 8 using the National Home Energy Rating scale (approximating to a Standard Assessment (SAP) rating of 80 see chapter 3) would be required to satisfy the affordable warmth criteria. A calculation done for a similar house in York with a SAP rating of 83 after modernisation (Bell et al. 1992) predicted an internal temperature of 18.5°C at a weekly average cost of £8.60. This compares with 18.5°C and £9 per week calculated in the example case used by Boardman (1993). Measurements of actual temperatures and energy consumption in the York house over a 12 month period give a whole house mean temperature for the heating season of 17.3°C and an average weekly fuel cost over the 12 month period of £7.25, representing some 7 per cent of the occupant's estimated income (Bell and Lowe 1995, see also chapter 8).

The requirements for affordable warmth set out by Boardman are commensurate with the energy standard defined by the Energy Rating Method used for the first time in the 1995 Building Regulations (a SAP rating of

75

between 80 and 85 depending on floor area). Given the current rates of new house building however, changes to building regulations are unlikely to deliver affordable warmth in the foreseeable future. Significant improvements in the existing stock will therefore be required and as we demonstrate in the case study in chapter 8, the required levels are attainable in many existing dwellings at reasonable cost.

Boardman's model is contingent upon factors such as the price of fuel as well as income and the efficiency of the dwelling. Changes in any of these factors could make fuel poverty worse or better. The precise level of affordable warmth is to a large extent a moving target and there is, undoubtedly, room for discussion about the level at which affordability and warmth should be set. In addition it can also be argued that by focusing the debate on energy, Boardman draws attention away from the underlying problems of poverty and income distribution. Despite these qualifications, the concept of affordable warmth provides a clear and useful model which can be used to address one of the major issues in energy and social policy.

Existing temperatures in English homes

Few large scale surveys of internal temperatures in houses have been undertaken in the UK. A survey of room temperatures was however included in the English House Condition Survey for the first time in 1986 (DoE 1991).[2] Although based on spot readings this gives a useful picture of temperatures attained across the country. Table 6.1 sets out house mean and living room temperatures for mean external temperatures of 2°C and 0°C. As one would expect, living room temperatures were generally higher than the house mean. If a whole house mean temperature of around 18°C is used as a bench-mark for warmth (Boardman 1991), these figures suggest that only about 28 per cent of households achieve such a level when the external temperature is 2°C, falling to around 25 per cent at 0°C. Taking a combination of house mean and living room temperatures, only around 10 per cent of homes met recommended design temperatures (21°C living room 18°C other rooms - BS 5449 1990) at the time of interviews (there is no significant change as the temperature falls). At the cold end of the range, about half of English dwellings were kept at temperatures of less than 16°C with a slight rise (from 49 to 53 per cent) as the temperature fell. Perhaps the most striking change however is that the proportion of houses with a whole house temperature below 9°C rose markedly (from 6.4 to 9.9 per cent) as the external temperature fell to 0°C.

Because these temperatures are based spot temperatures they cannot be directly compared with the whole house mean temperature levels advocated for warmth. It is likely that since the temperatures were taken at the time of interview (during the day and evenings) they would be in the upper region of the 24 hour cycle for the houses surveyed and thus whole house average temperatures would be lower. The picture presented by these figures is a rather bleak one, suggesting that there is a long way to go in achieving the temperature levels required for comfort and for health.

Table 6.1
Temperatures in English homes (1986)

Temperature range	Home mean temp. @ external temps.		Living room temp. @ external temps.	
°C	2°C	0°C	2°C	0°C
	(%)	(%)	(%)	(%)
Over 21°	4.5	3.9	13.1	12.8
18° - 21°	23.8	21.3	35.6	33.4
16° - 18°	22.7	21.7	27.2	21.2
12° - 16°	32.1	30.9	21.9	29.0
9° - 12°	10.5	12.3	2.2	3.6
Below 9°	6.4	9.9	-	-
Total	100.0	100.0	100.0	100.0

Source: DoE 1991

Energy efficiency and health

If a large part of the population cannot afford to maintain adequate temperatures, this not only results in concern about comfort and increased propensity to debt and disconnection, but also about health. The most widely discussed issue relates to the elderly and the concern about hypothermia and other cold related deaths. The broad evidence on hypothermia is derived from statistics of hypothermia deaths and external air temperatures which show the highest mortality rate occurring in the first quarter (January to March) when external temperatures are at their lowest (Collins 1993). In addition, however, concern has also been raised about the relationship between cold and damp homes and other medical conditions particularly cardiovascular and respiratory problems.

Cold related deaths

That there seems to be a significant relationship between cold and deaths either from hypothermia or other causes (heart disease, strokes, bronchitis and pneumonia) is broadly supported by national figures. In the case of hypothermia - taken as an underlying cause of death - the statistics show that most deaths occur in the coldest quarter (January to March). It is also evident that the variation in death rates in the first quarter taken from year to year are related to ambient temperatures. The figures for 1988 to 1990 are particularly striking with very mild external temperatures (around 7°C) and a marked fall in deaths Collins (1993). Although deaths from hypothermia are highlighted in the media they represent only about 1 per cent of all cold related deaths. The extent of the more general problem of cold related deaths, is indicated by the number of deaths which occur in winter, compared with deaths at other times of the year (referred to as excess winter deaths). Markus (1993) presents figures for Scotland which range from an average of 5,577 in the 5 year period 1958-62 (42.1 per cent) to 3,468 in the period 1983-87 (24.5 per cent). Within the last 20 years or so the number of excess winter deaths in the UK has varied between 59,000 in 1976 and 25,000 in 1987 (Boardman 1991a).

The existence of an association between external temperature and death rates has been established for some time. Using data for the years 1963 to 1966, Bull and Morton (1978) were able to show a close association for most diseases other than cancers. In a more detailed study of deaths from myocardial infarction, strokes and pneumonia (data from 1970 to 71), Bull and Morton (1978) also demonstrated that a rise in the external temperature from -5°C to 17°C (England and Wales) was strongly associated with a fall in the death rate and that the fall was linear over the full temperature range. They also showed that for heart problems and strokes, the slope of the regression line was much steeper in the case of those over the age of 60 than for those less than 60. A comparison with data from New York shows a similar picture within the temperature range -10°C and 20°C. However, the regression of deaths on temperature is less steep in the USA data than that for England and Wales. In their discussion of this comparison, Bull and Morton (1987) suggest that warmer internal temperatures in New York may explain the difference. They also point out (citing work by Anderson and Le Richie (1970)) that in Toronto where good control of internal temperatures exists, there is no significant correlation between death rates and myocardial infarction and strokes. We will return to international comparisons later in this chapter.

Given the age of the above data, (some 30 years) there is a need, perhaps, for a repeat study designed not only to update the general picture but also to attempt to relate it to changes in living conditions and other factors. However, the evidence of a continued high level of excess winter deaths would suggest that the general association between external temperature and death rates from most diseases continues.

Table 6.2
Blood pressure response of adults to temperature

Temperature	Response during an exposure of 4 hours at rest.
6°C	Increases in blood pressure found in both young and elderly subjects.
12°C	Blood pressure increases observed in older subjects after 1 hours, increasing further after 2 hours. No increases in young subjects.
15°C	No increase in blood pressure in either group of adults.

Source: Collins 1993 after Collins et al. 1985

In addition to statistical evidence, it is also important to be able to trace the physiological changes which accompany exposure to low temperatures. Work in this area has been reviewed by Collins (1993), Mant and Gray (1986) and by Raw (1988). In the case of cardiovascular responses to cold conditions, low temperatures can lead to changes in heart rate and blood pressure. Table 6.2 presents a summary of the findings from a study of adults (young and old) exposed to a range of temperatures for 4 hours at rest. In addition to blood pressure changes, the study also observed a fall in deep body temperature of 0.1°C in young subjects rising to 0.3°C in the elderly at 6°C. Continued exposure to 6°C (4 hours per day for 7 to 10 days) did not result in any adaptive change. Thus it seems unlikely that people exposed to the cold 'get used to it' in any physiological sense (Collins 1993 after Collins et al. 1985). In addition to increased blood pressure in response to the cold, changes also seem to take place in the composition of the blood. In a study of mild surface cooling in young adults, Keatinge et al. (1984) observed parallel increases in blood pressure and blood viscosity (21 per cent), as well as in platelet and red cell counts. The effect of these changes is thought to be an increased risk of arterial thrombosis since the increase in platelets and red blood cells promote platelet adhesion to blood vessel walls, which, combined with an increase in blood viscosity could lead to blood clot formation. Although the study did not include elderly subjects, the implications for this group (who have a higher level of initial arterial disease) would be an even greater risk of coronary and cerebral thrombosis. Keatinge et al. (1984) also suggest that the timing of the change in blood composition accords well with the time lag of roughly 24 hours between a reduction in temperature and an increase in death rate due to coronary thrombosis, which was calculated in the study by Bull and Morton (1987).

Despite the strong statistical and physiological evidence for a causal link between cold conditions and mortality, particularly with respect to deaths from coronary and cerebral thrombosis and respiratory disease, it is difficult to establish a clear causal link with the level of internal temperatures (as opposed to external temperatures). The first national study in 1972 of hypothermia and heating among the elderly summarised by Wicks (1983) (see Wicks 1978 for a full account) found that although the incidence of clinical hypothermia was very low, almost 10 per cent of the sample were 'at risk' (deep body temperature less than 35.5°C). The study also documented very low temperatures in both living-rooms and bedrooms. Only 29 per cent of living room temperatures achieved over 20°C by late afternoon (only 10 per cent of morning temperatures were in this class). Bedroom temperatures were particularly low with 4 per cent below 6°C, 33 per cent below 10°C and 84 per cent below 16°C. Notwithstanding these important findings, the study was not able to establish a significant correlation between body temperatures and cold homes. Similar findings emerged from a repeat study carried out by the Institute of Gerontology at Kings College London (Salvage 1993). Despite a small increase in temperatures (due to an increase in the prevalence of central heating - up from 24 per cent to 67 per cent), many of the elderly were still living at temperatures which are below the comfort level (81 per cent of morning living-room temperatures were below 20°C) (Salvage 1993).

The association between indoor temperatures and general levels of mortality is also difficult to disentangle. Using central heating as an indicator of warmer homes, differences in central heating provision (in 1982) across eight regions of England and in Wales did not lead to significant differences in seasonal mortality (Collins 1993 after Alderson 1985). However, one must question whether the variation in central heating between one region and another (a maximum variation of around 6 per cent from the average) is large enough to show any difference, especially when the availability of central heating does not necessarily mean that it is used to maintain adequate internal temperatures. In a further study (Keatinge 1986) which looked at mortality rates in elderly residents of warm sheltered housing, it was found that a rise in mortality took place in winter which was of a similar proportion to that in the population as a whole. The conclusion drawn from these studies is that external temperatures encountered on going out of doors may play a larger part in excess deaths than the level of indoor temperatures which currently prevail. In commenting on these findings Collins has remarked that:

> It does not explain why excess winter mortality is much less pronounced in countries with warmer homes and colder outside winter conditions unless there are large behavioural and clothing differences which prove to be critical. (Collins 1993)

It is perhaps the international comparative evidence which presents the strongest case for a significant contribution from low internal temperatures. We have already noted the comments of Bull and Morton (1978). Boardman's comparative analysis of seasonal mortality in a number of countries (Canada, France, Norway, Finland) reveals that, despite having the warmest winters, the British Isles also has the greatest seasonal variation in mortality. Boardman also endeavours to relate seasonal mortality to internal temperatures and despite difficulties in obtaining good comparative data, is able to demonstrate that countries with high internal temperatures have low seasonal mortality rates and that countries with lower internal temperatures (notably England, Wales and Scotland) display high seasonal mortality rates. The correlation between data items is significant at the 0.05 level. (Boardman 1991a).

A further study by Keatinge et al. (1989) looked at the growth in central heating ownership (from 13 per cent in 1964 to 66 per cent in 1984) and winter mortality. Even allowing for the varying coldness of winters, excess winter mortality from respiratory disease declined by 69 per cent over the period studied. Some of this fall may be explained by factors such as fewer epidemics of influenza, a fall in cigarette smoking and more effective antibiotics. Unlike respiratory disease however, there was no significant fall in excess winter mortality attributable to coronary disease or strokes. Since an increase in internal temperatures is associated with central heating ownership (Shorrock et al 1992), the study provides further evidence of the role of internal temperatures. Collins suggests that these findings shed further light on the debate surrounding the contributions made by internal and external temperatures to excess winter deaths.

The present evidence therefore suggests that changes in the respiratory mortality in the elderly in winter is most closely related to improvements in home heating and to the occurrence of influenza epidemics. The effects of outdoor cold may be more dominant in causing excess coronary and cerebrovascular deaths, though there may be an indoor factor if the home environment is particularly cold. (Collins 1993)

It is clear that cold conditions play an important part in mortality but the relative importance of external and internal conditions has not been established. Physiological evidence suggests causal mechanisms but is still inconclusive. There is undoubtedly scope for further work in this area, particularly in updating the various studies to reflect changes in the state of the housing stock over the last two decades. It is clear, however that if comfort temperatures are attained, the risk of strain on the cardiovascular and respiratory systems will almost certainly be reduced, with a consequent reduction in the risk of mortality. On the other hand, if people are living at temperatures such as those found in the study of the elderly (Wicks 1978) the physiological evidence would point to an increased risk, an important proportion of which must stem from internal temperatures.

Cold and damp homes and illness

Concern is often expressed about the effect of cold and damp homes on general health, particularly that of children. The evidence is again not clear cut and there is significant debate about the epidemiological evidence of a causal link between cold homes and illness. Hunt (1993) describes two studies, the first among a sample of the residents of 300 dwellings in Edinburgh and a larger study of housing in Edinburgh, Glasgow and London (a total of 597 dwellings covering 1124 adults and 1169 children, Hunt et al. 1988). Both studies adopted a 'double blind' methodology in that data on dampness was collected by environmental health staff independently of the collection of information on health problems from householders. The studies found that there was a strong relationship between dampness and health problems in children. The largest study (Hunt et al. 1988) identified a significant relationship with the health of adults although this was less marked than in the case of children's illness. Both studies implicated the presence of mould growth as well as dampness, although again it was children who had the largest range of symptoms showing a significant relationship. The principal symptoms included vomiting, wheeze, fever and poor appetite in children and nausea in adults as well as bad nerves and feeling low. A 'dose-response' relationship was identified for different levels of mould and for different levels of dampness. Results have also been reported in the United States which also show a strong and consistent relationship between home dampness and other non-chest illness (Brunekreef et al. 1989).

In her review of dampness, mould growth and health, Hunt (1993) identifies a range of agents which may damage health when damp conditions prevail. Infection by viruses seems to be more common in damp houses and damp conditions provide fertile ground for the growth of bacteria. Dust mite populations increase considerably in damp conditions (40 per cent or more relative humidity, Korsgaard 1979) and the debris they leave behind, particularly faecal pellets, have been strongly implicated as allergens leading to respiratory problems. The relatively pure water conditions produced by condensation are conducive to the growth of mould and the subsequent release of fungal spores. The implication of mould in both allergic and toxic reactions is strong and they may also interact giving an increased effect. In addition it is possible that the effects may be related to the vulnerability of the inhabitants of damp housing, particularly young children, the elderly and those suffering from other illness (Hunt 1993). In the study in Edinburgh, Glasgow and London referred to above (Hunt et al. 1988 and Hunt and Lewis 1988) some 404 moulds were identified. This study was not able to show a specific relationship between particular moulds and specific symptoms, despite a more general relationship. It was concluded that there were likely to be significant interactions between fungi and that the presence of many types of fungi, working together, might give rise to the symptoms observed. In pointing out the difficulties of identifying the exposure of individuals to mould Hunt argues that we do not yet have the methods to quantify all the variables which would be needed to establish a mean exposure to mould. She concludes:

Currently therefore, the health consequences of long-term exposure to mould in the home have not been precisely established in uncontrolled environments. Nevertheless, the repeated findings of associations between the presence of mould and symptoms of ill-health, together with evidence of clinical assays, leaves little reason to doubt that exposure to some fungi can constitute a significant health hazard. (Hunt 1993)

It is generally accepted that the problem of condensation and mould growth is minimised if a relative humidity (RH) below 70 per cent is maintained (Mant and Gray 1986). However since 40 per cent RH is associated with growth in dust mite populations, 70 per cent RH may not be an adequate level where dust mite debris may be a cause of allergy.

In contrast to the studies by Hunt and her colleagues, a study of damp housing and asthma (Strachan 1993) raises doubts about studies which rely (even in part) on self reporting of symptoms. In Strachan's work, although reported dampness was significantly correlated with reported symptoms, independent tests of both asthma and environmental conditions call into question the results based on reported symptoms.

[ ] attempts to validate reported symptoms and housing conditions by objective measurements cast doubt upon the reality of the association, and provide some indirect evidence of differential reporting, at least of respiratory symptoms. (Strachan 1993)

In a study of 297 children (Ross et al 1990) over a 6 month period in which measurements of temperature and humidity were taken and both health records and reported symptoms noted, it was concluded that:

No association between upper respiratory tract infection and domestic temperature or humidity levels could be shown in this study. (Ross et al. 1990)

Clearly some uncertainty remains despite very compelling evidence (supported by significant methodological innovations) presented by Hunt and others.

Indoor air quality and 'tight' houses

With the development of relatively air-tight designs for houses in a bid to reduce ventilation heat loss concerns arise as to the quality of indoor air. Work in this area is still in its infancy and there is much more to be done. A review of American evidence by Du Pont and Morrill (1989) indicates that the most important variable seems to be the strength of the source of pollutant rather than the tightness of the house. In one study (Nitschke et al. 1985) not only was no correlation found between air-tightness and indoor pollution but the tightest house had the lowest level of pollutant and the most leaky had the highest level. It was also found that houses with the highest concentrations of pollutant had significant internal sources such as kerosene heaters, smokers, wood stoves and fireplaces. It is clear that there is unlikely to be a simple relationship between air-tightness and indoor air quality. It may be that both tight and leaky houses could exhibit poor air quality depending on location and the existence of an internal or external pollutant source. Air change rates advocated for the control of condensation (between 0.5 and 1 air changes per hour) have been tentatively suggested as being reasonable to guard against the build up of indoor air pollutants (Mant and Gray 1986 see also BRE 1994 and the Building Regulations 1995).

Conclusions

It is clear from our review of the issues of poverty and health presented in this chapter that energy efficiency policy must demonstrate that it is tackling these problems, in addition to concentrating on environmental issues. Although the two objectives are unlikely to be in conflict, it should not be taken for granted that solving one will also solve the other. It would be quite possible to alleviate fuel poverty without significantly reducing carbon dioxide emission rates. Similarly energy consumption in dwellings could be reduced as a result of efficiency improvements but energy pricing policies could still lead to hardship for those in poverty. Tackling both objectives will almost certainly require related but different policies. In the realm of fuel poverty, Boardman's work provides a clear model which can be used in the development of policies to alleviate fuel poverty. However further work is required to establish ways in which the goal of affordable warmth can be achieved. It is clear that significant emphasis should be on the energy efficiency of housing for low-income households but additional income support may be required in the short to medium term which would help to alleviate some of the immediate problems.

Although there are some contradictions in the evidence on mortality and indoor temperatures, there is strong support for the view that the level of excess winter deaths (particularly among the elderly) is related to the level of internal temperature found in many dwellings in the UK. Similarly, the evidence of a link between cold and damp homes and illness is also strong, despite methodological reservations expressed by some researchers. The evidence is mounting that children are particularly at risk and that the presence of mould growth may be responsible for a number of illnesses. A resolution of the debate will require more detailed epidemiological studies as well as detailed work on the effects of mould growth.

The conclusions of research into the extent to which improved air-tightness resulting from energy efficiency measures affects indoor air quality suggests that the problem is more likely to be one of source strength rather than air-tightness. Air-tight houses do not seem to be any more prone to indoor air quality problems than leaky houses. However the work in this area is still in its infancy and conclusions must therefore be treated with some caution.

Notes

1 Although air temperature is used in this context, most practical instruments for measuring and controlling temperature in dwellings, respond to a mixture of air and radiant temperature.

2 Temperature data also formed part of the 1991 English House Condition Survey. At the time of writing, the energy report is still to be published.

7 The behavioural dimension

Introduction

It is self evident that because it is people (rather than houses) who use energy, consumption will be significantly influenced by behaviour and that the effectiveness of technological solutions will depend to a significant degree on how they are applied. Despite its importance, our understanding of behaviour, whether it be the decision to invest in energy efficiency or the use of heating controls, is rather sketchy. The headline figures in the literature are clear and it is common to see statements indicating a two to one variation in space heating for identical houses, much of which is attributed to behavioural differences. Sonderegger's classical analysis of consumption patterns in dwellings where changes in occupancy had occurred (movers) compared with dwellings in which there was no occupancy change (stayers) is often quoted in support of this assertion (Sonderegger 1978). Other references include Seligman et al. (1978), Scholes and Fothergill (1975) and Verhallen and Van Raaij (1981). Williams et al. (1985) cite further studies which demonstrate that large variations exist which cannot be explained by the physical characteristics of the dwelling.

While it is accepted that considerable variation exists in the way households use their home, it is important to look a little closer at what this means. The studies by Sonderegger (1978 - USA) and Verhallen and Van Raaij (1981 - the Netherlands) are particularly interesting in filling in some of the detail. In Sonderegger's study it was concluded, that some 54 per cent of variance between movers and stayers, was the result of physical differences and that 46 per cent was due to some other cause. The study was principally concerned with explaining the 46 per cent. Retaining the total variance as the base, the 46 per cent is broken down to 33 per cent occupant factors and 13 per cent non-obvious physical factors (standard of, or changes in construction). The 33 per cent splits into, 15 per cent associated with factors to do with family or income development (for example; new baby, wife goes out to work) and 18 per cent day-to-day behaviours. The work of Verhallen and Van Raaij reported 24 per cent of variance due to physical characteristics, 26 per cent due to reported household behaviour (temperature settings, use of curtains, airing out of rooms and use of bedrooms) and 11 per cent to special characteristics (working wife, illness and prolonged absence from the dwelling). All identified variables accounted for 58 per cent of total variance.

The above analysis shows that not all of the non-physical variation can be attributable to everyday behaviour. Sonderegger suggests 18 per cent, although this may not be representative of all types of households, given the (unavoidable) homogeneity of the study group. Changes in family size, age and the employment patterns of family members (15 per cent in Sonderegger's study), all play a part in establishing the differences in consumption but they are not things which a family can do much about. The suggestion from studies of this nature is that changes in behaviour could have an impact on energy efficiency and consumption. This is true up to a point but the scale of the contribution is not likely to be as large as the gross figures would indicate. To conclude therefore that a major improvement in energy efficiency can be brought about by changes in everyday behaviour alone, is no more tenable than the assertion that the solution lies solely with technology. For example, since many low income families are already living at sub-comfort temperatures, fuel poverty could not be alleviated by encouraging a change in behaviour (which would reduce temperatures even further), nor would it be sensible to advocate that the better-off should live at the same temperatures as the fuel poor. The primary point which needs to be emphasised is that the consumption of energy in an efficient way is influenced by physical, economic, social and behavioural factors acting in concert. This chapter deals with the behavioural and social issues in an attempt to draw a more complete picture.

Attitudes and energy efficiency

Opinion poll studies in the USA have shown considerable popular support for renewable sources of energy and energy efficiency over the last 20 years (Farhar 1993). In the UK, research into the success of media campaigns has indicated that 90 per cent believe that 'it is important to save energy' and 65 per cent feel there will be an energy problem in the future. Even so in 1979 only 1 per cent spontaneously mentioned energy as a problem facing the country (Gaskell and Pike 1983). A qualitative study of 103 people in 94 households in the UK also shows a positive attitude to energy and environmental issues (Hedges 1991). Although positive attitudes exist, there seems to be little understanding of the relationships and issues involved. Hedges (1991) points to the woolliness of many of the concepts and the fact that despite wishing to avoid damage to the environment, interviewees did not really know what to do about saving energy and often found it hard to believe that what they did at home would have much effect. Research into American public opinion suggests that the problem is seen in terms of social institutions rather than individual behaviour. As in the UK, qualitative interviews suggest that there is limited popular understanding of the issues (Lutzenhiser 1993).

Studies of general attitude and self reported energy conservation actions have consistently failed to find a clear relationship between the two variables (Olsen 1981 and Gordon et al. 1981). When related to actual energy conservation behaviour, the conclusion is broadly the same (Verhallen and Van Raaij 1981). In general, psychological research fails to show a clear relationship between attitudes and behaviour. The implications of this research is that, changing general beliefs is unlikely to induce a substantial change in behaviour. However, beliefs which relate to some variable which is of immediate personal concern may well influence actions. In the energy conservation context, this would mean that changing attitudes and beliefs about the impact of energy efficiency on a more salient concept such as comfort may change behaviour (Williams et al 1985). Evidence from the Twin Rivers study in the USA supports this view. In two surveys of summer energy consumption related to a range of attitude variables, the only set which showed a consistent relationship across both surveys were to do with comfort and health. Other factors (the likely cost and benefit and the importance of individual action in conservation) seemed influential in the first survey but not in the second (Seligman et al 1978). A further survey undertaken during the winter showed comfort and health to be important, along with an optimism that science would solve the energy problem. Other variable sets (legitimacy of the energy crisis, family finances and anticipated savings) showed no relationship with consumption. (Olsen 1981 after Seligman et al. 1979 and Becker et al. 1980).

Attempting to discover the prime influences on behaviour remains problematic. Williams and Crawshaw (1987) suggest that behaviour could be affected by changes in attitudes which relate to salient factors such as comfort or by reconciling apparently conflicting beliefs, such as the belief that comfort and saving energy are incompatible. In a further study (Black et al. 1985), it was found that (in line with other studies) generalised concern for energy did not influence behaviour directly but that an indirect influence was exerted through an effect on personal norms, such as a feeling of personal duty to conserve energy. The set of personal norms most amenable to this effect are those related to curtailment of consumption, rather than to improvements in energy efficiency. The influence of external constraints was clearly identified as a factor, in that the fewer impediments to a given energy saving action which existed, the more likely a person would be to act on personal norms. Hence, simple curtailment of consumption such as switching off heating in certain rooms or lowering temperature settings, were more amenable to personal norm influences largely because these actions were less constrained (and, perhaps, undertaken with less conscious deliberation) than actions involving an investment in energy efficiency works. Black et al. (1985) concluded that this combination of external constraints and the intervening cognitive processes relating to personal norms, probably accounts for the failure of studies to find a correlation between general attitudes and energy saving actions.

This research would suggest that little faith can be placed in media campaigns which present general messages aimed at altering attitudes but which have no specific personal relevance. Even after almost 20 years of exhortation and debate on the energy question, Hedges (1991) is able to report that people are broadly in favour of energy efficiency but do not know how to achieve it and are concerned that comfort should not be sacrificed. The notion of energy literacy has been identified as an important element in energy efficiency behaviours (Gaskell and Ellis 1981 and Gaskell and Pike 1983). It is quite possible for people to believe that they are being energy efficient when an objective assessment would show that they are not. For example, people may state that they are very careful in the use of energy, but waste heat through an unlagged hot water tank (Williams et al. 1985 after Green and Ventris 1983). The clear message in this context is that policy must seek to take into account the beliefs, understandings and motivation on which behaviours are based, together with the constraints under which people operate, rather than rely on changing general attitudes. Some of the questions which need to be addressed are summed up in the following quotation:

The principal contention [ ] is that efforts to conserve energy, or to use it more efficiently, need to concentrate much more on the behaviour of the consumer, how he uses his environment and how he changes his strategy in response to economic forces. In particular more knowledge is needed on cognitive aspects of household behaviour. How does a person achieve his aims be they to conserve energy or simply to keep warm? And how does his behaviour relate to his budget or to his particular type of dwelling? (Williams et al. 1985).

Motivation and energy efficiency

Although the evidence is clear that energy consumption is influenced by behaviour and that this can account for a significant variation in consumption, the driving force behind those behaviours is very unclear. The work on attitudes indicates the strength of concern for comfort and health (Seligman et al. 1978), but does not really say a great deal about what actions are regarded as producing appropriate levels of comfort and what such actions mean for energy efficiency. It is quite likely that beliefs about energy efficiency are overridden by concerns about comfort and health (Williams and Crawshaw 1987). The majority of people in one study (Crawshaw et al. 1985) over-ventilated their homes and just over half did not associate opening windows with a possible increase in heating bills. In her study of energy efficiency and the elderly, Salvage (1992) also noted a concern with maintaining 'fresh air' with 41 per cent of respondents reporting that they kept bedroom windows open in the winter and 70 per cent feeling that it was important to open windows all year round. The belief in fresh air would appear to be a strong one but is one which is in conflict with the goal of energy efficiency. Only one respondent spontaneously made the connection between window opening and the costs of heating. Work on air quality in naturally ventilated bedrooms suggests, however, that opening windows, particularly when sleeping, is a reasonable thing to do so as to avoid the build up of carbon dioxide concentrations (Feist et al. 1994). The critical question is; what represents the right balance between ventilation needs and energy conservation needs?

Attempts to motivate people to save energy, either through efficiency improvements or simple reductions in consumption, have stressed the need to change attitudes or to use the price mechanism. As we have seen, changing behaviours by seeking to change general attitudes is unlikely, of itself, to be successful. However there is some hope that a deeper understanding of the complex attitude and belief pathways which exist, would help in identifying avenues of approach which could support knowledge and information strategies (Black et al. 1985). The use of the price mechanism as a motivating force also presents difficulties. It would appear that consumers are more likely to adapt to price rises (assuming that they are able) and that they continue to consume at similar levels despite an initial tendency to cut back (Gaskell and Ellis 1982). Research by Dillman et al. (1983) concludes that, in the face of price rises, low income families make cut backs in all areas of use (to the detriment of their life style) but that higher income households maintain their consumption, as well as taking advantage of any incentives available to invest in energy efficiency. Within the confines of the current climate it is likely that the price mechanism is more likely to increase fuel poverty than reduce consumption. In their Massachusetts study Black et al. (1985) concluded that energy price increases were unlikely to result in investment to reduce consumption or to persuade people to make ambient temperature adjustments and may be more effective in bringing about hardship rather than saving energy.

These insights into the response to price, need to be seen in the context of our discussion of long run energy prices and energy taxation in chapter 4. Historically energy price rises have not been sustained, the sharp rises of 1974 and 1979 each being followed by a fall (see figure 4.1). Over the period since 1987 prices have fallen steadily, with a particularly large fall in the price of the main heating fuel, natural gas. Thus, in our view, any attempt to measure the long term response of the domestic sector to price increases against this background is likely to fail. This does not mean, however, that the effect of a sustained and carefully introduced price rise would be negligible. In attempting to understand such an effect, there are at least two issues which need to be explored. One concerns the internal dynamics of the way in which physical improvements to the housing stock are made and the other, concerns the beliefs and expectations of owners and occupiers.

To consider the question of system dynamics; where any system is forced to respond more quickly than its own internal dynamics allow, it is likely to produce transient behaviour which may run counter to the main objectives of any long term change. In the case of energy prices, the main responses to a rapid and unplanned price rise are likely to be the sort of deprivational and price absorbtion behaviours discussed above, as well contributing to general inflation. The time taken for the domestic sector to respond fully to a price rise is determined, however, by the decades long maintenance cycle in the dwelling stock and the even longer cycles involved in stock replacement. It is likely that when energy prices change slowly, perhaps over a period of ten years or more, and predictably, that significant energy efficiencies will be induced. If prices rise rapidly and then fall, as has been the case in the last 20 to 25 years, the incentive to improve or build to higher standards is dissipated. By the time owners and occupiers have begun to consider investments in efficiency measures, the force of the economic case has been significantly reduced.

On the question of the beliefs and expectations of owners and occupiers, the relationships involved are undoubtedly complex and any response to price changes, however introduced, will be mediated by the cognitive and social mechanisms discussed elsewhere in this chapter. In the context of the current discussion it would seem likely that any response to price rises will depend not only on current price, but on beliefs (and hopes) about future price levels. If one's experience suggests that by sitting tight it is possible to ride out a price rise, the incentive to overcome the considerable barriers to investing in energy efficiency will be greatly reduced. In fact such an approach is potentially more economically efficient, particularly if the duration of any price rise is short. The implications, for energy policy, of the question of price are important. In order to harness the potential of price through a policy of energy taxation (a policy we discuss in chapter 4), government would need to take into account the ability of the domestic sector to respond through investments in efficiency and address existing expectations and beliefs derived from the past which would undermine such a policy.

Given that it is difficult to discern a straightforward relationship between attitudes and motivation or price and motivation, it is useful to look at the process of motivation itself. In this context the expectancy theory of motivation in management science provides some important insights (for an explanation of this theory see Robertson and Smith 1985). The theory postulates that motivation is dependent on an expectation that something (say energy efficiency) can be achieved and is within the remit of personal action (expectancy), that actions can be identified which will achieve a desired outcome (instrumentality) and that the outcome is valued or is considered to be worthwhile given the effort required (valence). The insight provided by this theory is potentially useful, in that it deals with the processes involved in motivation rather than the more general theories of motivation which deal with issues of self or human needs. From the above discussion of attitudes it would appear that energy efficiency is valued but that expectancy and instrumentality are low. In addition, it may be that, in a climate of falling real fuel prices, valence may also be low. There is a lack of knowledge about what level of energy efficiency is possible and what needs to be done to achieve it (Williams and Crawshaw 1987). Stern and Aronson (1984; after Stern et al. 1981 & 1982) provide an interesting example of what happens when some of these gaps are filled. In an energy efficiency programme which consisted of energy audit, finance, contract management and inspection, 2000 households (about 25 per cent of those requesting audits) took advantage of the full package. The reasons they gave for participating were; trust that works would be inspected (98 per cent), no worry in finding a reliable contractor (96 per cent), trust in the professionalism of the staff (89 per cent) and convenience (62 per cent). Even where people did not have the full package, the provision of information through the energy audit (possibly filling the instrumentality gap for those individuals), induced activity which would save a large amount of energy. Very few households mentioned finance as an important reason for participating in the programme.

Tonn and Berry (1985) in their development of a mathematical model of decision making in energy conservation, suggest that people seem to adopt a 'pruning approach' to decision making which selects the least risky or simplest to understand course of action, prior to considering such things as pay back. In fact they conclude that pay back may not be a significant factor at all in retrofit decision making. An analysis of implicit discount rates, suggest rates which are extremely high (over 60 per cent in many cases). Such high rates imply considerable reluctance to invest and this reluctance is largely associated with the 'hassle' and uncertainty factors involved (Williams R H 1989). Further evidence of the 'hassle' factor is provided by Boardman (1991b) referring to the take up of draughtproofing works on a Sutton Housing Trust estate in Manchester. About 25 per cent of tenants on benefit (for whom the work was free) refused to have it done because they did not want the disturbance. This problem with respect to draughtproofing was also recognised by the research on attitudes done by Hedges (1991). The motivation to be more energy efficient clearly needs to address the cognition issues involved in bridging the gap between general attitudes and action.

At the same time, those variables which are intrinsic to a household's situation will also influence energy efficiency behaviour. These are often referred to as contextual variables and usually create significant practical barriers to making improvements. As discussed in chapter 4, tenure has an effect on improvements. In the case of rented property there is a general lack of incentive to invest on the part of both the tenant and the landlord. Black et al. (1985) in their path analysis of factors which determine capital investment, identified home ownership as the single most important influence. The influence of ownership has also been referred to by a number of other researchers (Gaskell and Pike 1983, Salvage 1992, Williams and Crawshaw 1987). In the case of social housing, the benefits of energy efficiency are not only associated with lower fuel bills for the tenant but can be experienced by the landlord in indirect ways (see chapter 4). Changing the perception that energy efficiency is not worth a landlord's while because the pay back accrues to the tenant, could play an important part in landlord motivation. This will require further demonstration and quantification of the benefits not only in the public but also in the private sector.

Access to capital for improvements is a major obstacle for many households and for those in fuel poverty, an insuperable one. The net effect is that those on higher incomes can obtain access to capital to enable them to reduce energy bills while the only options available to those on low incomes is to curtail energy use or to seek aid through government schemes such as the Home Energy Efficiency Scheme. Access to capital is, however, not only a function of household income but also of the attitude of government and the lending institutions (Stern and Aronson 1984). The nature of these contextual barriers is just as much part of the overall behavioural picture as are the barriers created by such issues as lack of knowledge and understanding.

Knowledge and understanding

The state of knowledge

The qualitative work by Hedges (1991) indicated considerable uncertainty and confusion about the whole area of energy efficiency in the home and the actions which could be taken to achieve improvements. The importance of knowledge and understanding has been identified by a number of researchers (Blumstein et al. 1980, Cook and Berrenberg 1981, Gaskell and Ellis 1982, Gaskell and Pike 1983, Williams and Crawshaw 1986, Williams and Crawshaw 1987, Crawshaw et al. 1985, Salvage 1992, Salvage 1993). In a study of 160 households in London, Gaskell and Pike (1983) showed a significant relationship between knowledge and gas consumption (significant at the 0.05 level and accounting for 5 per cent of total variance).

The issue of householder knowledge as a barrier to energy efficiency has been reviewed by Salvage (1992) and her review, supported by qualitative interviews with 100 elderly people in London, reveals a significant lack of knowledge among consumers. Understanding of the benefits of investing in energy efficiency is low, as is general knowledge of the efficient operation of heating controls and the running costs of appliances. Most suggestions for fuel saving measures in the interviews carried out by Salvage (1992) were essentially 'deprivational' rather than energy efficiency ones. The tendency to deprive, rather than to improve efficiency, is of major concern given the issues of fuel poverty and health discussed in chapter 6. This tendency is also observed in other studies across a wider range of household types. Even where efficiency improvements were identified for householders in an audit, savings resulting from curtailment and behavioural changes were consistently overestimated when compared with savings from efficiency measures such as insulation and heating improvements (Lutzenhiser 1993). Crawshaw et al. (1985) show that specific knowledge can be critical. Even where there was general understanding of the relative running costs of heating and appliances, a lack of specific knowledge showed a correlation with higher bills. The ability of consumers to understand energy consumption is exacerbated by the nature of fuel bills which aggregate costs and are in any case difficult to understand even at the level of checking the calculation (Stern and Aronson 1984). It has been demonstrated (Williams 1989) that a redesigned fuel bill could increase understanding considerably and give consumers more control over their use of energy. In Sweden a scheme has been developed (started in

1989) using an 'Energy Letter' which is sent to customers giving details of electricity consumption over the recent past and containing suggestions as to how consumption can be reduced. Savings of 12 per cent have been reported with zero capital investment (CADDET 1994). Similar measures have been undertaken in Denmark (Danish Ministry of Energy 1990)

The provision of knowledge is a complex task since the needs of consumers are extremely varied and as Crawshaw et al. (1985) have demonstrated, highly specific knowledge can be required. However, people develop their own simplified methods to account for consumption using money as the prime unit. Such variables as running time or the amount of human labour replaced by an appliance are used to estimate the consumption of equipment. Because the methods used produce systematic errors, actions taken or decisions made about energy efficiency are often far from optimal in energy terms. It would however be wrong to dismiss the methods as just inaccurate, since the methods used say a great deal about the way in which complex information is handled. Kempton and Montgomery (1982) have coined the term 'folk model' to describe this process of energy quantification. The folk model is cognitively efficient in that it is easy to use, easy to learn and enables comparisons with other purchasing activity such as buying food or clothes (Kempton and Montgomery 1982). The implications of this insight for policy making are important, in that it points to the possibility of harnessing the model in such a way as to correct the errors it produces.

Providing knowledge, feedback and advice

A number of studies have attempted to assess the effect on energy consumption of providing feedback and/or information to consumers. Environmental psychologists have been interested in the analysis and modification of behaviour related to the use of resources. Research in this area suggests that information given before an event (antecedent information), is less effective than the adoption of a feedback strategy in which information on consumption is provided as consumption proceeds (Geller et al 1982). However the effectiveness of antecedent information (or prompts) can be enhanced by careful design of the communication (Lutzenhiser 1993, Winett et al. 1982). Williams and Crawshaw (1987) conclude that prompts can be effective if actions are clearly specified, do not involve a great deal of personal effort and are not perceived as likely to reduce comfort.

The number of studies involving feedback is large but despite this, the picture is still a rather murky one. A comparison of a number of studies carried out by the same research team will serve to demonstrate some of the difficulties involved. Seligman and Darley (1977) reported a significant reduction in summer electricity consumption (much of it related to air conditioning usage) of 10.5 per cent following feedback based on each consumer's past consumption. A second study (Seligman et al. 1978) reported similar results when feedback was given combined with the setting of a difficult savings target (a 20 per cent reduction). However a feedback group with an easy target and two no-feedback groups (easy and hard targets) showed no significant reduction in energy consumption. A third study, also reported by Seligman et al. (1978), involved a light to signal when the air-conditioning was not needed but was running (a signal to turn it off). Households were allocated to light plus feedback, feedback only, light only and no treatment (control) groups. Home owners with the light (feedback and no-feedback) reduced their consumption significantly (15.7 per cent) the other groups (feedback only, no treatment) did not. The conclusion was that the feedback had no effect.

Unravelling this, we see that the nature of the feedback and the way messages are given are important in understanding the ways in which people act. The explanation given by Seligman et al. was that subjects in the feedback only group in the third study, disregarded the information because it 'jumped around too much' and was not thought credible. What Seligman et al. fail to point out, however, is that the blue light also constituted feedback. This was the most potent feedback of all, because it was immediate, provided the simplest message ('if it is on, turn off your air-conditioning') and had credibility.

Another study of summer cooling (Winett et al. 1977) looked at financial incentives and feedback, concluding that feedback was effective and that the incentive component was only significant if it was a large one. It is also interesting, given the comments about the strength of concern for comfort above, that the feedback was not effective on warm days. The effectiveness of providing trained advisors giving highly specific advice on the use of air-conditioning and summer domestic water heating was also tested by Winett et al. (1982). Their results although based on small samples (N=14, 13 and 12 for each sub group) suggest a beneficial effect on consumption with reductions of around 21 per cent for groups receiving advice.

Although these studies (Seligman and Darley 1977, Seligman et al. 1978 and Winett et al. 1977 & 1982) provide useful insight into the role of feedback, they all relate to summer consumption which is dominated by air-conditioning and care is required in their interpretation with respect to winter heating, which is the main concern in the UK. A study of winter consumption in the UK by Gaskell and Pike (1983) cast doubt on the value of feedback on its own. However an information strategy which involved detailed discussions with occupants of ways in which energy could be conserved, as well as written information, resulted in significant reductions of 15.5 per cent in gas and 9.5 per cent in electricity consumption. These reductions did not depend on whether information only or information plus feedback was given. Of critical importance was the provision of information.

Given that many feedback studies involved daily feedback, the sustainability of savings once the research projects had been completed can also be a problem (Lutzenhiser 1993) and monitoring systems or additional incentives may need to be developed to maintain initial savings (Cook and Barrenberg 1981). Gaskell and Pike (1983) report evidence from their follow up study (after 12 weeks) which indicates that the information effects were relatively long term. Williams and Crawshaw (1978) suggest that:

> In general feedback appears successful if presented frequently (at least twice a week), in written form, in a cumulative time scale, based on one's own rather than someone else's past consumption. [ ] Carefully presented feedback can help in energy management. But evidence is needed to identify the most effective channels of communication and the most useful dimension to represent the information. (Williams and Crawshaw 1987)

Stern and Aronson (1984) identify the importance of imitation and interpersonal contacts as powerful forces in energy behaviour. They point out that in both day-to-day consumption and decisions about efficiency investments, the informal networks of friends, family and colleagues are influential. The use of a friend's experience with a particular energy improvement, household appliance or method of operating a heating system, is likely to have greater meaning and saliency than other types of information and is also more likely to be trusted. A comparison between the effectiveness of local community groups and an energy audit company in encouraging the take up of energy audits demonstrated a much higher success rate (15 per cent) in the case of the community group than the professional company (6 per cent) (Stern 1986 after Polich 1984). The reasons for this, it is argued, relate to uncertainty and a lack of trust in certain formal information sources.

The credibility of the information source would appear to be of particular importance. Stern (1986) discusses a number of studies which illustrate its importance. In one study of energy advice (Craig and McCann 1978), a brochure on how to reduce energy was sent to households in New York. Half were sent under the heading of the New York State Public Service Commission and the other half under the banner of the local electricity company. In the following month a 7 per cent saving in consumption was made by the first group but no savings were made by the second. A similar picture emerges in the more constrained area of investment in energy efficiency improvements. A study of a shared savings programme (Stern 1986 after Miller and Ford 1985), in which the cost of energy improvements were recovered through energy savings, used three different communication approaches. One group received a letter from the private company who were managing the scheme, a second group received the same letter but which mentioned that the local authority (Hennepin County, Minnesota, USA) were involved and the final group received a similar letter but on the note paper of the local authority and signed by the chairman of the county Board of Commissioners. The response from the three groups is set out in table 7.1. The results clearly show, by a factor of 5, the pivotal role played by the local authority in encouraging participation in the scheme.

Table 7.1
Minnesota shared energy savings study

	Company letter only	Company letter with county backing	County letter
	(%)	(%)	(%)
Request for audit	6	11	31
Signed up to the works	1.7	2.7	9.3

Source: Stern 1986 after Miller and Ford 1985

The need for energy advice has been recognised in a number of public sector refurbishment schemes in the UK. In one scheme (EEO/WS Atkins & Partners 1986), an advice programme of monthly visits over the space of nine months resulted in reductions in consumption, increased use of central heating, improved control of heating, reduced condensation and improved comfort when compared with households who received no advice. A scheme involving a group of 10 tenants trained as Neighbourhood Heating Advisers (NHA) was evaluated (Optima Energy Services 1989) and concluded that central heating systems were better understood and used more effectively by tenants on the estate with advisers, compared with tenants on an estate where no advisers operated. These improvements were reported despite the fact that less than a quarter of households received advice from an NHA. The study also concluded that although the approach seemed to be an effective advice delivery mechanism, its full potential was probably not realised in the scheme. Realising this potential raised issues of management of the NHAs and the extent to which volunteers could be called upon to take a more active role.

The results of advice provided as part of an improvement scheme to housing association houses in Merseyside, again demonstrated the impact of advice to tenants (Tong 1987). The advice programme included visits from a field worker, advice sheets and an energy calendar containing energy tips and a space for recording consumption. A reduction of 10 per cent (after allowing for temperature changes) in energy consumption was observed during the winter in which advice was provided. It is not clear if this result was considered significant but changes in behaviour were reported which centred on reduced use of immersion heaters and changing control settings to reduce room temperatures.

Although the advice projects reported beneficial effects, the extent to which they can operate outside the confines of a local authority or housing association housing scheme are uncertain, and more work is required on such issues as delivery methods, saliency of information and quantification of savings.

Design of systems

Although this chapter is primarily concerned with occupant behaviour, it is important to assess the relationship between occupant behaviour and the design of technological systems. This is an obvious area of work particularly given the growth in ergonomics since the middle of this century. However this area seems to have been neglected by ergonomists. Evidence from advice studies such as those indicated above and from research into the state of householder knowledge, suggest that many people experience difficulty with heating controls. Lutzenhiser (1993) and Salvage (1992) cite a large range of studies which illustrate a low level of understanding of control systems and behaviours such as the use of thermostats as on/off switches. There is nothing to say of course that this sort of manual operation is less efficient than automatic operation but it could have important consequences for system design.

The need for an integrated approach which takes into account both the technical and human aspects of operating a house, has been identified at a number of levels, ranging from the design of heating programmers to whole house design (Williams et al. 1985, Williams and Crawshaw 1987, and Jesch et al. 1987). Exploratory work by Salvage (1993) on heating controls for the elderly made a number of recommendations concerning clarity of display, ease of manipulation and consideration of problems caused by poor memory among the elderly. It was also recommended that the needs of those who will enter old age in the next 10 to 15 years should be taken into account in developing control systems which relate to their understandings of technology. Despite the obvious importance of understanding human-house interactions, the field does not appear to be well covered when it comes to design guidance.

Social processes and energy efficiency

There is a clear consensus in the literature that an understanding of energy and behaviour must include an understanding of the social context of individual action (Lutzenhiser 1993). These influences can be seen in a number of areas. Stern and Aronson (1984) identify role models and the power of trusted members of immediate social circles as influences on behaviour and on the knowledge which informs that behaviour. Darley and Beniger (1981) demonstrate the impact of social networks in the diffusion of energy efficiency technology. The power of social pressures to reduce consumption can be seen even in situations (for example, master metering in blocks of flats) where the individual, acting as an individual, and on purely economic criteria, would benefit from increased consumption (Lutzenhiser 1993).

Hackett and Lutzenhiser (1991) argue that energy consumption is to some extent 'built in' to social identities and that within those identities consumption is 'obligatory'. So much so that when people from other cultures enter an existing setting there are a number of social pressures which oblige them to change behaviours so as to match the consumption associated with their new social identity. Homes and their intrinsic energy characteristics are also powerful expressions of membership of, and status within, a community and society (Stern and Aronson 1984). Appliances must conform to status expectations and energy efficiency is only one of many issues in social settings (Lutzenhiser 1993). To ignore the idea that energy efficiency must be congruent with the social context is to miss an important piece of the jigsaw which makes up the totality of domestic and non domestic energy consumption. As an area of research Lutzenhiser remarks that:

> The social organisation of consumption within families and communities - whether treated as a network phenomenon or as institutionalized group behaviour - is a potentially fruitful, but largely overlooked, area of analysis for behavioral energy research. (Lutzenhiser 1993).

The cultural dimension is also an important one, but one which raises large questions about cultural and sub-cultural life styles (Lutzenhiser 1992, & 1993). Comparative studies across national and continental boundaries reveal a large number of differences in the use of energy in general and energy efficiency in housing in particular (Schipper and Ketoff 1985 and Schipper et al. 1985). The integration of social and cultural research into existing models of energy consumption and efficiency, holds some promise for greater understanding but would appear to be still in its infancy. Perhaps the most important aspect of culture in modern industrialised societies which bears on the energy debate, is the deeply embedded expectation of material growth. This is a factor, the importance of which extends beyond the confines of a book on housing but which, in the long term, cannot be ignored.

The behaviour of organisations

Households and organisations live in the same world and the social structures overlap. Boardman (1991a) has made the point that many of the lowest income groups do not get the benefit of 'free heat' during the day because they do not go to work, a fact that exacerbates fuel poverty. Stern and Aronson (1984) identify organisations as intermediaries who shape energy efficiency behaviour by limiting choice and availability with respect to such things as electrical appliances, building insulation or heating systems. They argue that intermediaries have few incentives to pursue energy efficiency since they will not reap the consequences of their actions. Anecdotal evidence from the York energy demonstration project (Bell and Lowe 1995, see also chapter 8) would suggest that the concerns of contractors in sizing condensing boilers in well insulated houses had nothing to do with the energy efficiency of the boiler or the efficiency with which the capital was used. Stern and Aronson (1984) point out that although the consequences of energy consumption are felt by the user, it is not the user who is in a position to put market pressure on manufacturers. Since the review by Stern and Aronson in the early 1980s, Lutzenhiser (1993) notes that little work on organisation behaviours and energy efficiency has been carried out. It should be pointed out however that this does not apply to the technological and management issues where considerable developments in energy management systems are taking place under the EEO's best practice programme (see for example Eclipse Research 1993a and 1993b and BRECSU - OPET 1992).

The role of government as an intermediary is a very important one. Quite apart from their general legislative function, they can provide (as was illustrated above) a degree of credibility to the way in which energy advice or improvement schemes are communicated. Other organisations may also play such a role. The legislative function with respect to the construction of new housing and the design of domestic appliances, is also crucial to any energy efficiency strategy. Building Regulations and codes of practice (a type of intermediary) are in a pivotal position and play a large part in the standard of efficiency which is achieved in both new build and renovation projects. Unless there is a push from the market, house builders in the private sector are unlikely to either want higher energy standards or build to higher standards voluntarily (Davies and Pyle 1993). The government's role, as a provider of housing funds, can be used to influence the behaviour of those organisations which make use of such funds. For example, the standards of energy efficiency in the new building and modernisation schemes of housing associations are to a significant degree set by government through the funding mechanism.

Knowledge of the way in which intermediaries influence energy efficiency is critical, since it is they who design new housing, or modernisation schemes. Building in energy efficiency is always more cost effective at times of major change. Research by Blumstein et al. in the USA over 15 years ago (Blumstein et al. 1980) indicated a rather low level of knowledge, not only among households but also among building professionals. There are signs that knowledge is improving but a recent survey of UK professionals indicated that although environmental and energy issues were beginning to feature as one of their concerns, only some 12 per cent placed it first out of a list of 5 construction industry priorities. Perhaps of more concern however, is the fact that in the same survey only 3 per cent of clients rated energy and the environment the most important issue (Building 1995).

Understanding of the relations between consumer and producer and the impact of intermediaries in the field of energy efficiency is, to say the least, sketchy. A reliance on models which assume that producers only strive to fulfil consumer demands and emphasise individual choice and action or rationality is likely to be a dangerous course (Lutzenhiser 1993).

Models of energy consumption

The formulation of policy designed to improve energy efficiency and reduce consumption will depend on how the problem is viewed. If a designer is asked to design a new building, the result will depend on the designer's previous experience. To present a caricature; a steel designer is likely to design a steel building, a concrete designer a concrete building and a timber designer a timber building. The real solution to the client's problem however, may not be a building at all (a point made by Koehler 1987). To use another analogy; Stern (1986) opens a discussion of what he sees as 'blind spots' in economic theory as applied to energy use, with the story of the drunk and the lamp post. The drunk was looking for his keys under the lamp post, because that was where the light was! The obvious problem illustrated here is that although different disciplines shine a light onto the field of energy use, they are often isolated spots of light (of varying brightness), none of which shine on the lost keys. The message to be drawn, is that in order to search the ground more fully, it is necessary to explore a range of view points and the spaces which lie in between. A failure to do this is likely to result in ineffective policy formulation.

An additional difficulty for policy making is that the models used are often implicit rather than explicit. For example, calls for grant or loan schemes depend on a rather simple view of the way people respond to price, which in turn, is based on a classical economic model of behaviour. Very often, however the basis of the model is not stated. Although it could not be argued that price is unimportant, as anyone who has worked in private sector housing improvement will testify, simply making renovation grants available (even at 90 per cent) does not guarantee the successful improvement of an area. Energy use (as with housing renovation) is, of course, much more complex than a simple economic formulation of the problem will allow. It is also more complex than any formulation which has a single focus. It is important for policy making to be able to see the underlying models which are being used and to treat them as ways of shedding some light on the problem, rather than as complete and self-contained theories. Lutzenhiser (1992) identifies four broad classes of model (Engineering, Economic, Psychological and Social/Anthropological) and their characteristics are discussed below.

Engineering models

Engineering models are concerned primarily with the hardware of energy consumption and its efficiency. Simulation models such as BREDEM, are able to make accurate predictions of energy consumption given closely defined parameters. The physical parameters, such as level of insulation and heating system efficiency, are relatively stable but the user parameters (by comparison) are not. The engineering model does not claim to understand fully human actions or motivation and the human in the system is treated largely as a 'body in motion' which affects heat gain through metabolic processes and desired heating periods (Lutzenhiser 1992). This is not to say that technologists do not appreciate the other issues or models involved, but rather, that the engineering model cannot understand these human aspects and can only make assumptions about behaviour patterns and plug them into the physical models. The factors which drive those patterns are not the concern of the model itself. The danger of using a purely engineering based model, is its overemphasis on physical solutions without taking into account their interactions with users.

When attempts are made to take users into account, there is also a danger that they are treated (perhaps subliminally) in a similar way to the technology itself. Although there is some evidence to support the view that people are generally positive about technology in the home (see, for example the response to the idea of home automation in RMDP/NEDO 1989), it is clear that it is often not understood, particularly when it comes to heating controls. The danger in this situation lies in the engineering model's response to such problems. In order to remove the instability created by the user's lack of understanding, it is tempting to remove control from him or her. In certain circumstances there may be some justification for this, but removing control can often be counter productive, as well as raising philosophical and ethical problems. There is some evidence from studies in office buildings, that thermal comfort is, in part, related to the extent of control users have over their environment and that reducing control options produces a progressively narrower comfort band requiring tighter and tighter engineering (and more and more energy) to maintain (Baker and Standeven 1995). If only engineering solutions are sought, there is a danger of setting up a vicious circle in which more engineering simply leads to more engineering. The problem, as Williams et al. (1985) argue, is that not enough thought is given to the way both physical and human systems combine to produce a more desirable total system outcome.

To develop another line of argument, the engineering model has little to say about the way in which technological developments are incorporated into professional practice and applied in design and modernisation work. This is another sort of user problem which relates to the understanding of those who design and construct the technology. As the drive to produce more efficient devices and more efficient structures has intensified, large reductions in consumption are possible with no reduction in comfort (an increase in comfort in some cases), yet we are still discussing the problems of take-up and application. It has been pointed out (Bell and Lowe 1993) that reliable cavity wall insulation has been available for over 20 years, yet resistance to fully filled cavities still exists among some building professionals. At the construction end of the chain of application, there is concern about installation quality. For example, there is evidence that calculated heat loss characteristics (U values) of certain wall constructions maybe optimistic by as much as a factor of two (Siviour 1994). The problem of accurate calculation and prediction is essentially an engineering problem but ensuring that the required level of insulation is achieved in practice is fundamentally a human one.

Economic models

Lutzenhiser (1992) argues strongly that the principal limitation of economic models is their adherence to the notion of rationality among consumers. In other words, the notion that in any given set of circumstances, it can be assumed that people will act according to a set of reasoned actions designed to maximise some utility function. To pursue this line of argument is, however, not only difficult but also rather sterile, in that it all depends on how rationality is defined. If it is defined according to some automaton-like construct with a limited set of objective variables, it is easy to show that humans do not behave in a rational way. If, on the other hand, it is defined as a decision process which leads to actions which an individual believes to be right in all the circumstances, as that individual knows them, then all humans are rational in their own terms. If, given the latter definition, it is concluded that our models are not good descriptors or predictors of economic behaviour, we cannot blame the concept of rationality, all we can do is recognise that we do not fully understand human behaviour.

Economic models with their emphasis on response to price and the notion of discount rates in personal investment decisions, provide useful models for some aspects of energy efficiency. However, as we have discussed above, the link between price and behaviour is not a simple one. Stern (1986) points out, that a reliance on economic models, brings with it the danger that the problem is seen in economic terms only and leads to blind spots when comes to the formulation of policy. Stern's analysis seeks to show that there are difficulties with the theories of price and investment in understanding energy consumption, that there are important determinants of behaviour which economic models ignore and that the dynamics of economic behaviour are not well understood. Although economists would not agree with his broad conclusions on the underlying theory of price and investment, the remaining two concerns would seem to be acknowledged (Quigley 1986).

Any response to a movement in price is determined not only by the change itself but also by the extent to which it is perceived by householders (an issue of attention, Stern 1986) and other non economic factors. Economists seek measures of price elasticity to understand the relationship between price and demand but this does not say much about the underlying behavioural processes. We have already observed that the state of knowledge about energy consumption is generally low and that providing knowledge either as 'energy tips' or as feedback could be successful if the information were well designed. Policy action which addresses the information issue may be more effective in changing consumption behaviour than adjusting the price. In simplistic terms, if the price goes up but consumers do not understand how to reduce energy consumption (other than by forgoing significant thermal comfort), the effect is likely to be small. In addition, where people are unable to afford to respond by paying more, the result may well be greater fuel poverty and, perhaps, political pressure to reduce the price or mitigate the effects of the price rise through the benefits system. Taking a more holistic view however, a more coordinated approach which used the price mechanism (perhaps using taxation to ensure that prices did not fall in real terms) combined with adequate information and other support systems may be even more effective.

Purchases of energy and energy consuming devices are not only investments but have cultural meanings outside the economic sphere (Stern and Aronson 1984). An understanding of these cultural meanings and the way they influence investment decisions is probably just as important as assessing pay back times. Similarly, an understanding of the role of 'ease of decision factors' (Tonn and Berry 1985), or the effect of external constraints on investment behaviour (Black et al. 1985), are as relevant to policy as implicit discount rates. The fact that implicit discount rates are high may not reflect a short sighted time preference of consumers but the difficulties of organising the works, or the uncertainties about what investment to make, or both. The role of intermediaries in the process and their often conflicting interests add further to the complexities of the relationships involved.

Psychological models

Psychological models are implicit in much of the material reviewed in this chapter. In particular the models relating to attitude. As already observed, general attitude is not a very good predictor of behaviour and this leads to a view which rejects this as an approach to understanding energy efficiency behaviour (Lutzenhiser 1992). However evidence is presented in the work of Black et al. (1985) that an understanding of attitudes and norms is an important part of the picture. Psychological models which are concerned with the provision of information and the way in which users understand their energy use, suggest closer links with actual behaviours, but on the evidence presented in this review they have not developed much beyond identifying that a significant lack of knowledge exists and that well presented information and feedback can result in reductions in consumption. A model based on the human-machine system concept in ergonomics (Sanders and Mc Cormick 1993) has been proposed by Williams et al. (1985) and at a practical level represents a sharp focus on detailed behaviour. The problem with this model is that it can be seen (although it is rarely presented in this way) as little more than an information flow model with exchanges between the human and the house (a machine in this context), rather like driving a car or flying an aircraft. Some elements of energy efficiency and behaviour are like this, as in the design of heating system controls, but many are not. However the model can also be used to draw in other psychological aspects as well as the economic, social and technological issues which affect the relationship between people and the houses they occupy. Despite the fact that the model has been around almost as long as the study of ergonomics itself, it does not seem to have received much attention as a potential force for improving our understanding of energy behaviours or the development of practical policies.

Sociological\anthropological models

Lutzenhiser (1992) argues that although psychological models have acknowledged a social dimension as it impinges on the individual, they do not account for the dynamic behaviour of human groups themselves. This concern with the role of social groups and their place in the settings with which domestic energy efficiency is concerned is a more recent addition to the energy efficiency literature. The extent to which social structures and the cultures from which they stem, influence the behaviour of individuals making decisions to insulate or turn down the thermostat or to accept their lot and be cold, is not clear and there is still a long way to go in modelling the factors and links involved. Models of social behaviour also raise hard questions about the sharing of energy between groups in society in an equitable way. Much of the concern with fuel poverty is of this nature. The existence of fuel poverty when there are obvious technical solutions which would make many people better off while not making others worse off (the classical Paretian optimum (Winch 1971)) requires an understanding of the social forces which operate either to bring about change or to prevent it.

It is not the intention of this discussion of models to imply that there is a single line of approach. The different model sets represent different aspects of what is a multi-faceted problem. Some of the areas involved and some of the linkages are identified in figure 7.1. Rather than argue for one aspect or another (an argument that will often split on professional lines) it will be necessary to look for ways in which they combine to present a more complete picture which can be used to inform policy. The goal is, after all, improved total system performance which leads to a sustainable level of energy use. It is our contention that this cannot be achieved by looking for solutions in one area only.

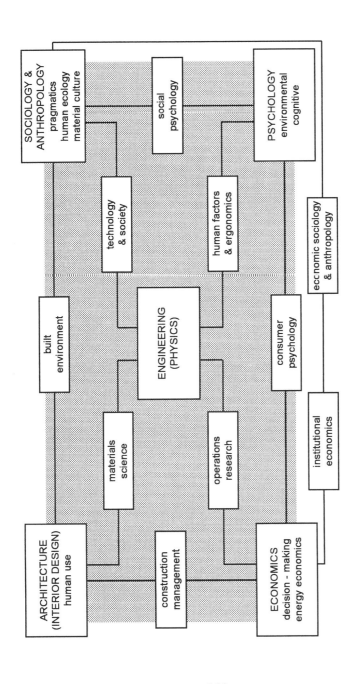

Figure 7.1 Disciplinary specialities and concepts relevant to the study of human/environment and human/technology relations

Source :   *Adapted from Lutzenhiser 1992*

109

Conclusions

The literature reviewed in this chapter leads to a number of broad conclusions:

1. Energy efficiency and consumption is significantly influenced by household behaviour but that behaviour is part of a system which has behavioural, social and technical dimensions. The result is a function of all three acting in concert and effective policies need to be based on an understanding of the relationships involved.

2. General attitudes to energy problems do not have a direct effect on energy efficiency actions although they may have an indirect effect on behaviour (whether to invest in improvements or to cut back on consumption) through other more specific concerns. In general, attention needs to be focused on these more specific concerns such as the need for comfort and health and to demonstrate that energy efficiency is compatible with such requirements. Personal norms which see energy conservation as an individual responsibility may result in energy curtailment actions but are unlikely to lead to energy efficiency investment.

3. The motivation to invest in improvements is generally low and people are more likely to choose measures which are unlikely to involve disruption. The concept of economic pay back does not appear to be a strong influence in investment decisions. It appears that people are more likely to choose measures perceived to involve the least uncertainty and risk, rather than those which provide the best pay back. There is little work which assesses intervention mechanisms designed to remove some of the disturbance or 'hassle' factors.

4. Knowledge and understanding are important requirements for behaviour and general levels of knowledge are low. This applies to the understanding of where energy is consumed in the home as well as the more specific understanding of how to operate controls or fuel bills. The knowledge required is often very specific and if provided at a general level is unlikely to be effective.

5. There is evidence that the provision of advice and information can lead to lower energy consumption both in existing situations and if provided within the context of an improvement scheme. This may or may not involve feedback information on past consumption. Feedback on its own is unlikely to lead to energy reductions. Our understanding of information

and feedback and the best way of providing it is still rather unclear and more work could usefully be done in this area. In particular, there are avenues such as the use of fuel bills and fuel meters which could be explored as feedback and information mechanisms.

6. The design of house energy systems particularly heating controls need to relate much more to the needs of householders. The elderly are a special group in this regard. A Human-Machine Systems approach would seem an appropriate one in this respect.

7. Social processes exert an influence on both energy consumption and the propensity to invest in energy efficiency. Much of the literature is still at a general theoretical level and there are many questions which need to be explored with respect to energy efficiency.

8. Knowledge of the part played by intermediaries is also still at a general theoretical level. In view of the identification of the energy utilities as emerging players in energy efficiency (see chapter 4) an assessment of their role could be important.

9. Energy efficiency or the lack of it is the result of a system with different facets. People are more than a source of heat gain, they do not act on economic criteria alone, they have models about how their world works and base actions on those models, they also operate within a social framework and at the same time they consume energy. Our understanding of the processes involved is far from complete and research is needed which attempts to deal with this complexity in such a way as to provide a sound basis for practical policies.

8 Case study 1: The York Energy Demonstration Project

Introduction

Improving energy efficiency in housing will depend on actions taken to improve the design of new housing and to adapt and modernise the existing stock. This chapter and the next are designed to provide some practical examples of the sort of improvements which are possible using existing, well established technology. This chapter deals with the practical issues associated with the improvement of existing housing. Improving the existing housing stock presents a unique set of opportunities and problems. Unlike the design of new housing, modernisation is constrained by the existing structure and fabric, existing orientation and other site factors, conservation issues and the specific needs of individual occupiers such as the requirement for continued occupation during works. These constraints have often led to the view that major improvements in energy efficiency are unlikely to result from existing stock improvements and that new-build is the most desirable option. While it is true that the potential of existing housing is not so great as in new-build, it is also true that significant improvements can be made during modernisation works and that it is possible to reduce energy use by as much as 50 per cent in much of the stock without the need for wholesale rebuilding. It is in any case not a very practical proposition to expect the sort of rapid increase in stock replacement which would be required to tackle the environmental and social problems of fuel poverty and health within a reasonable time scale.

To a large extent, improving the efficiency of existing housing is a matter of taking opportunities when they present themselves. It is, for example, much more cost effective to install double glazing when window frames are in need of replacement on repair grounds than to replace them on energy grounds alone. There are exceptions to this however. For example, cavity wall insulation does not usually require other repair works unless there are problems with the cavity itself and is, therefore, reasonably independent of other improvement works. The critical factor in determining the cost of cavity wall insulation, as we have already observed in chapter 4, is the size and type of contract. The case study presented in this chapter seeks to demonstrate the nature and scale of the contribution which can be made by improving existing housing as well as some of the practical problems which need to be solved.

Energy and existing housing

Housing accounts for just under 30 per cent of national energy consumption and a similar fraction of carbon dioxide production (DTI 1992). Given this fraction, it is clearly important that housing modernisation policies take account of energy and the need to improve energy efficiency. Many projects over the last 20 to 30 years have concentrated on developing the design of new housing and improving building regulations (see for example. DoE 1981, Everett et al 1985). However, in the UK, dwellings have a long physical life. Since 1970, demolition rates have declined and are probably running at about 20,000 per year, some 0.1 per cent of the stock. The long term annual rate of new construction is around 200,000 or 0.9 per cent of the stock (DoE 1991). Dwellings built after 1990 will probably constitute about 8 per cent of the total by the year 2000, and less than 40 per cent by 2050. To a large extent, new construction adds to the existing dwelling stock, rather than replacing it. Thus reductions in total domestic energy consumption will only come about if radical improvements in the performance of new dwellings are coupled with similar improvements in the performance of the existing stock.

Figure 8.1 illustrates the considerable variation of energy efficiency standard which exists within the British housing stock. Although such an analysis, based on age, is crude, it is nevertheless indicative of the order of magnitude involved, with a potential space heating reduction in the region of 70% in a typical 1930s semi-detached house improved to 1995 Building Regulations standard. Attempts over the last 20 years to improve the efficiency of the national housing stock have concentrated, with success, on loft insulation and lagging of hot water storage cylinders. By 1989 some 89 per cent of accessible lofts had been insulated with an average depth of 85mm and 90 per cent of cylinders lagged. Other important areas, notably masonry cavity walls and windows have also seen some improvement, with about 20 per cent of cavities insulated and 46 per cent of the stock receiving some double glazing by 1989 (Shorrock et al. 1992). In the case of double glazing however, very few houses are likely to be fitted with modern high performance double glazing units. Since these units have almost double the insulation performance of older units the potential for further improvement in glazing (even with existing commercially available technology) is probably of the order of 80 per cent.

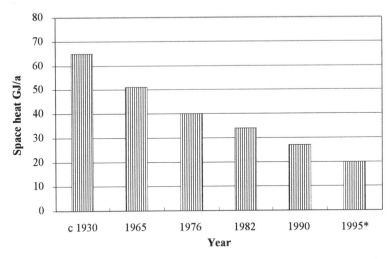

* Given that the 1995 level was not defined in the original analysis, this consumption is an indicative figure based on a similar house with an energy rating which approximates to the 1995 building regulations level (Bell and Lowe 1995).

Figure 8.1   **Impact of the Building Regulations on energy use (semi with well controlled central heating)**
Source: Adapted from Anderson (1988) crown copyright

The case study described in this chapter, addresses the energy conservation opportunities which present themselves in housing modernisation programmes. The results from the project, suggest that this approach can unlock a large potential within the existing stock for reductions in energy consumption and carbon dioxide emissions.

The York Energy Demonstration Project (YEDP)

In 1991 the York City Council received funding from the UK government to embark on an energy demonstration project, the objective of which was to demonstrate the potential impact of energy efficiency measures on the modernisation and improvement of existing low-rise housing. This project was one of a number of demonstration projects set up under the UK Government's Greenhouse Programme (DoE 1993b and 1994b), which was designed to provide energy efficiency models for the future modernisation and improvement of local authority and other social rented housing. In common with other Greenhouse demonstration projects, the York Project concentrated on well tried and tested methods of improving energy efficiency, the emphasis being on replication of well understood technology. The only exception to this was a heat-pump heating and ventilation system in one of the houses. The project consists of three individual schemes, each with a different emphasis.

In order to identify and disseminate lessons from the project, each scheme was monitored over a one to two year period. The results of the monitoring are summarised in this chapter and the implications for the wider application of energy efficiency in housing modernisation and maintenance are discussed. A more detailed account is presented in Bell and Lowe (1995).

The York housing stock and design of schemes

The Local Authority housing stock in York consists mainly of two storey semi-detached and terraced housing of traditional construction based on brick cavity walls and pitched tile and slate roofs. The City Council owns some 9,600 dwellings, which represents about 25 per cent of the total housing stock in York. Some 75 per cent (7,280 dwellings) of the council stock was built prior to 1964 and although the data in Figure 8.1 would suggest that there is much that could be done to all ages of property, the project concentrated on the pre-1964 stock, since this forms York's current modernisation target group. Although the project was aimed at the local authority housing stock many of the houses in the project area are typical of house types in the private sector in York. The measures demonstrated are, therefore, of clear relevance outside the local authority sector.

An assessment of energy conservation options, and the design of coherent packages of measures were carried out for each scheme in the project using the National Home Energy Rating programs (NHER) produced by the National Energy Foundation (for discussion of a precursor to these programs, see Chapman, 1990). The programs provided energy and cost estimates based on standard use patterns. All assessments assumed a dwelling occupied by 2 adults and 2 children, maintaining a living room temperature of 21°C for 9 hours per day Monday to Friday (2 heating periods per day), with a single 13 hour heating period at weekends. The schemes are described below.

Figure 8.2   4 house scheme: typical house

The 4 house scheme

Figure 8.2 illustrates one of the houses from this scheme. The 4 house scheme was designed to investigate a small number of properties in detail. It involved insulation improvements to a standard some 20 per cent to 30 per cent higher than the 1990 building regulations (the then current standard) and explored 4 different heating strategies, two gas systems and two electric. The systems were as follows:

- Gas House A was fitted with a full central heating system for space and water heating which included a condensing boiler with an efficiency of between 85 per cent and 90 per cent (some 20 per cent to 25 per cent greater than a conventional boiler).

- Gas House B had a system which used three gas room heaters, two downstairs and one upstairs. Water heating was provided by an instantaneous gas heater.

116

- Electric House A used an off-peak central heating system based on a dry-core storage boiler linked to a conventional radiator system. Water heating was provided by a conventional off-peak storage cylinder fitted with 75mm of insulation. A mechanical ventilation system with heat recovery was also fitted in this house.

- Electric House B was the most experimental of the 4 houses. A warm air system was fitted which combined heat recovery ventilation with an air to air heat pump. Auxiliary heating was provided by resistance heaters in the ventilation supply ducts and a focal point electric fire in the lounge.

The insulation measures consisted of cavity wall insulation, loft insulation (200mm.) and replacement timber windows with low-emissivity double glazing and draught seals. Works to improve the air-tightness of the houses were also carried out where appropriate. Monitoring of this scheme was done in two stages. Stage one was done on a before-and-after basis in order to provide a measure of the effectiveness of the improvements. This involved the direct measurement of the heat loss characteristics of the house fabric. Stage two consisted of long term monitoring designed to assess the performance of the houses during occupation. Internal and external temperatures and energy consumption were monitored over a twelve month period after the works and in some cases energy consumption was disaggregated to enable the effectiveness of heating systems to be understood in more detail.

The 30 house scheme

The objective of this scheme was to demonstrate the implementation of energy efficiency within an existing modernisation programme. Improvements consisted of wall insulation, loft insulation (200mm.) and draught proofing to existing doors and windows. No glazing improvements were carried out. The level of insulation achieved was some 10 per cent below 1990 building regulations (the then current standard). Gas wet central heating and hot water systems were installed using a condensing boiler. Monitoring of this scheme was designed to test the difference between existing modernisation standards and the enhanced standard achieved in the 30 houses. In order to do this, a control group of 20 houses was established which were drawn from the same modernisation scheme but improved to the normal modernisation standard used in York at the time. This standard provided for the refitting of kitchens and bathrooms together with minor alterations and repairs, but did not include any energy efficiency works other than the installation of a central heating system with a conventional boiler. This scheme presented a good opportunity to obtain a statistically convincing result based on actual consumption data. Monitoring consisted of internal temperature monitoring and gross energy consumption in each house. A typical house from this scheme is shown in Figure 8.3

Figure 8.3   30 house scheme: typical house

The 200 house scheme

This scheme was carried out some twelve months after the 4 and 30 house schemes and sought to apply the 4 house standard to a full modernisation scheme. Tenants had a choice of a gas or electric heating system. Most tenants chose a conventional gas central heating system with a condensing boiler. The scheme was monitored using a sample of 12 houses in which internal temperatures and energy consumption were recorded over a 12 month period. In addition to being able to assess the performance of these houses against the other schemes, an attempt was made to assess the impact of simple advice to tenants covering a number of aspects of energy efficiency in the home. Halfway through the 1994/95 winter a simple advice sheet was given to 8 out of the 12 monitored properties backed up by a personal visit from a member of the monitoring team. Energy monitoring then continued to the end of the winter and the results assessed.

Summary of results

The 4 house scheme

The short term monitoring of the 4 houses indicated that the insulation and air-tightness improvements had a significant impact on heat loss. The specific heat loss measurements, made before and after the works, showed a reduction in the region of 40 per cent. Long term monitoring demonstrated that three of the 4 houses performed largely as predicted but one of the electric houses (the most experimental which was fitted with an air to air heat pump) did not. This was mainly because the heat pump did not operate for large periods and the efficiency gains associated with it were not realised.

It was not possible to measure energy use in these houses before renovation and therefore this had to be calculated. The reductions in energy use were derived by comparing the calculated figures with the measurements of energy use after improvement. Table 8.1 sets out the results of this comparison.

Table 8.1
Delivered energy May 92 - May 93

	before improvement (kWh)			*after improvement (kWh)*		
	gas	elec.	total	gas	elec.	total
Gas House A condensing boiler	23900	4300	28200	13160	1209	13369
Gas House B gas unit heaters	23900	4300	28200	11535	1524	12059
Elec. House A off-peak elec.	n/a	24800	24800	n/a	12225	12225
Elec. House B air-air elec. heat pump	n/a	24800	24800	n/a	12296	12296

Source: Bell and Lowe 1995

The saving of delivered energy on this basis is about 14000 kWh averaged over the 4 schemes (15500 kWh, gas and 12500 kWh, electricity) representing just over 50 per cent. The potential cost saving at current energy

119

prices and tariff structures, would be in the region of £500. The exception to this is Electric House B where a large proportion of the electricity was consumed at the peak rate (63 per cent compared with 25 per cent in Electric House A) and the expected efficiency gains from the heat pump (which would have offset the higher rate tariff) did not materialise. The net effect was an energy cost saving closer to £240.

The internal temperatures, after improvement, in Gas Houses A and B (17.3° and 16.9°C respectively) are in line with data from other YEDP houses. The temperature in Electric House A (19.6°C) is high by UK standards, and compares with a mean of 18.4 °C for the Pennyland houses (Lowe et al. 1985) at roughly the same level of insulation. All of these temperatures are broadly in line with the affordable warmth standards discussed in chapter 6. Given the age and nature of the houses prior to improvement and the material on house temperatures presented in chapter 6, it is likely that the temperatures before improvement were lower and that a substantial proportion of the energy benefit was taken in the form of greater thermal comfort

In order to get an indication of the likely cash benefit, a direct comparison was made between the gas houses and the houses in the 30 house scheme control group.[1] The difference in the gas energy cost was around £175 (some 11,000 kWh)[2] and the additional capital cost for the condensing boiler scheme was just over £1000, giving a simple pay back period of between 5 and 6 years. In the case of the gas unit heater scheme the heating system was cheaper (around £1770 compared with £2200) than the central heating systems installed in normal modernisation. If this saving were set against the insulation improvements (about £760), a net cost of some £330 emerges giving a simple pay back in the region of 2 years.

Carbon dioxide emissions were reduced by about 4 tonnes per year in the gas houses (from 8 to 4 tonnes) and by about 9 tonnes in the electric houses (19 to 9 tonnes). The difference between the two fuels is a function of the large carbon dioxide overhead associated with the conversion of primary fuel to electrical energy in power stations. In comparison with the 30 house control group, the gas houses achieved a reduction of more than 2.4 tonnes per house per year (based on gas use only, see note 2).

The 30 house scheme

Although access difficulties and equipment failures reduced the numbers of monitored houses to 21 in the experimental group and 11 in the control group, there is a clear and significant difference between the energy consumption in the two groups. One of the more striking findings was that both groups maintained very similar internal temperatures over the heating season. A small difference of about 0.5°C in mean internal temperature was, however, observed. The average internal winter temperature for the experimental group was 17.9°C and for the control group 17.4°C. This difference amounts to something in the region of 1400 kWh (about £20 worth of gas) over a full heating season in these houses. Figure 8.4 shows average internal temperatures against external temperatures for each group. The scatter of the data points illustrate the closeness of the temperatures, even when external temperatures were at their lowest.

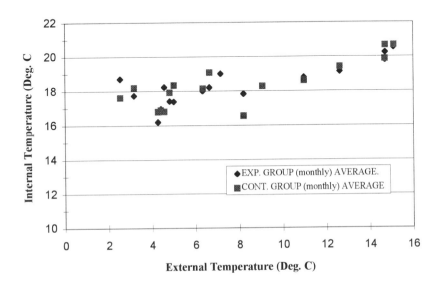

Figure 8.4   Internal temperatures plotted against external temperatures for experimental and control groups in the 30 house scheme

Source: Bell and Low 1995

The difference in average energy consumption between the two groups is set out in Table 8.2. Since all houses were heated using gas, one would expect improved insulation and the greater efficiency of the condensing boiler to result in reduced gas consumption. We observed such a difference which is statistically significant. The probability of this difference occurring purely by chance is less than 3 per cent (P=0.022). The difference in electricity consumption is not significant (P=0.201). Although a significant difference in gas consumption was observed, it was just less than half of the 11,100 kWh (about £180 in cost terms) which was predicted by the modelling programs under standard occupancy conditions.

Table 8.2
Comparison of measured energy consumption - 30 house scheme

Group	Gas		Electricity		Totals	
	kWh	Cost £	kWh	Cost £	kWh	Cost £
Experimental	19313	348	3138	292	22451	640
Control[*]	24458	430	3529	322	27987	752
Difference	5145	82	391	30	5536	112

[*] Consumption in control houses has been adjusted by +1366 kWh to allow for the difference in mean internal temperature.

Source: Bell and Lowe 1995

Tenant survey Tenant perceptions were surveyed before and after improvements in experimental and control groups. In both groups tenants felt warmer after modernisation and displayed high levels of satisfaction with their heating systems. This accords well with the temperature monitoring which showed that average temperatures were within comfort standards across both groups. The main areas of difference were that the number of tenants reporting condensation problems had reduced in the experimental group but not in the control group. In addition, the level of concern with heating bills was reduced in the experimental group but remained unchanged in the control group. Further analysis of the results from the 30 house scheme raised a number of important issues, some of which are discussed below.

Cost and pay back issues The extra cost of energy efficiency work on the 30 house scheme was £1442 (average), this would give a simple pay back of about 17 years based on the measured difference (temperature corrected) between the two groups. The predicted pay back was just under half this figure at 8 years. The costs for most of the house types in this scheme were higher than those obtained in the 4 house scheme. This was mainly because wall constructions were a mixture of cavity, timber and solid brickwork. Each required different treatment. The latter required dry-lining treatments which were more expensive than cavity filling. The 30 houses also incurred draught stripping costs of £182 per house which were not incurred as a direct cost in the 4 house scheme where draught stripping was incorporated into the replacement window frames.

Figure 8.5   House type C

An analysis of costs for house type C (the most complex and expensive house type) will serve to illustrate the cost issues involved. Figure 8.5 shows the house type and Figure 8.6 illustrates the simple pay back for each measure installed. In interpreting these results it is important to bear in mind that the impact of a particular measure depends on the other measures which are applied and the order of application. Generally speaking, the later an insulation measure is applied, the greater its effect will be. In the case of heating system efficiency the opposite is true. Putting a condensing boiler into a poorly insulated house will save more energy than the same boiler in a well insulated house. The measures evaluated for house type C were:

- Cavity wall insulation (cavity)
- Dry lining to the kitchen/utility room wall (DL - Util. rm.)
- Dry lining to the bay window wall and roof (DL - Bay)
- Insulation of the timber mansard roof (Mansard)
- Draught proofing of the windows (D proof)
- Installation of a condensing boiler (C boiler)

(abbreviations in brackets are those used in Figure 8.6)

123

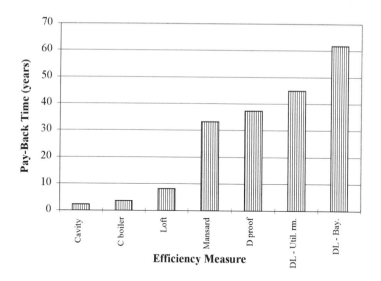

Figure 8.6   Pay back analysis for house type C
Source: Bell and Lowe 1995

In this analysis, the energy saving is calculated assuming that the measure is applied last. This means that insulation measures will show their maximum saving but the condensing boiler will show a lower saving than if applied with no insulation improvements.

The most effective improvements were cavity wall insulation, loft insulation and the condensing boiler. If dry-lining and draught proofing works were omitted the calculated pay back would be in the region of 4.5 years and if applied pro-rata to the measured difference the pay back would be around 8 years. These figures are in line with the pay back on the 4 house scheme where the wall insulation consisted of cavity fill only.

On energy grounds alone there would be little financial justification for certain additional insulation works in some of the house types in this scheme. However there are a number of comfort and amenity issues for the tenant which should not be overlooked. Omitting insulation in some walls or parts of walls is likely to create cold spots which will be prone to condensation and reduce the feeling of warmth for anyone sitting near to the wall. The omission of draught proofing may also reduce feelings of comfort in some rooms (see the discussion of comfort in chapter 6). The net effect of this may be to encourage occupants to raise air temperatures to compensate for lower radiant temperatures and higher air speeds and to combat condensation. This would, in turn, increase energy consumption. Experience would also suggest that as contractors get used to the house types and larger contracts are formed for certain items of work, costs would come down. For example, in the case of cavity wall insulation, costs have fallen from an average in the region of £150 per house in the 30 house scheme, to nearer £100 per house on subsequent (larger) cavity wall insulation contracts.

Use issues As discussed in chapter 7, large variations in energy consumption between houses of the same level of efficiency are common. The fact that in this case the difference between the two groups is not as large as the modelling programs predicted comes as no surprise. It seems reasonable to assume that use factors have tended to obscure the full effects of the energy efficiency measures. Although detailed investigation of use was outside the scope of the monitoring work undertaken, an attempt was made to assess in a qualitative way the likely issues. To this end a detailed interview was carried out with the occupants of one of the experimental houses.

The house exhibited an energy consumption some 40 per cent above the predictions of the modelling program despite internal temperatures similar to those which were predicted by the model. Part of the discrepancy would appear to relate to the use of the gas fire. The fire chosen by the tenant was an enclosed gas flame fire with an efficiency which varies between 59 per cent at high output and 47 per cent at minimum output. The use pattern which emerged during the interview indicated that the gas fire was operating on its low setting from about 2 o-clock in the afternoon to 11 o-clock in the evening and was also on for about an hour in the morning. The gas fire was used even during the timed heating periods. Since the lounge radiator has a thermostatic radiator valve, the heat from the gas fire would turn the radiator off for long periods especially during mild weather. This means that most of the lounge heat would be provided by the fire running at about 47 per cent efficiency compared with the condensing boiler at about 90 per cent efficiency. A crude assessment suggests that this could account for just under half the difference between measured and predicted levels of consumption. Similar fires were used in all houses in both experimental and control groups. The worst performance was achieved by a fire which had a claimed efficiency of 42 per cent at all settings. It also had an open flue which would have resulted in significant energy loss through over-ventilation, even when the fire was turned off.

The impact of gas fire usage in the two groups is likely to have a more marked effect in the experimental group than control group. This is because the discrepancy in efficiencies between primary (boiler) and secondary (fire) heat sources is greater in the experimental group. For example if a gas fire was providing 20 per cent of heating the increase in consumption would be 14 per cent in the experimental group but only 4 per cent in the control group. It would appear that both the choice of fire (assuming one is to be installed at all) and the way it is used are serious issues for energy efficient modernisation. Other aspects of use in the case investigated, related to the opening of windows in bed rooms hot water consumption and thermostat settings, most of which would tend to increase consumption in this particular case.

Where mixed heating systems are to be installed there is clearly a need to consider the range of fires offered and to provide guidance on both the choice of fire (if any) and its use, particularly in houses fitted with condensing boilers. This need for advice is further reinforced by the findings of the tenant survey, which showed that a very high proportion of tenants (80 per cent), used some combination of gas fire and central heating.

The 200 house scheme

Energy data was available for 10 houses (9 gas heated and 1 electric) in this scheme. The mean total delivered energy use in the gas heated houses was just under 18,600 kWh/a. Heating season internal temperatures could be reliably estimated in only 5 of the 10 and the median of these was 18.8°C, a figure which is slightly higher than in the other groups in the project as a whole. In comparison with the other schemes, the 200 house scheme falls midway between the 30 house scheme and the 4 house scheme. Although the efficiency measures applied in the 200 house scheme was broadly similar to that of the 4 house scheme there were important differences. Some houses had one wall which was not of cavity construction and funds were not available to fully insulate it. All gas houses had a gas fire (see discussion above) and because of tenant choice requirements, it was necessary to fit a non condensing boiler (with optimiser) in some houses. Given these differences the result is unsurprising.

Although tenant advice sessions took place during the last winter of monitoring, insufficient data was available to reach any meaningful conclusions as to its impact. However advice sessions provided some interesting insights, which are summarised below:

- Understanding of the heating system was mixed. Some tenants clearly understood the main points, some did not and often got relatives to set the time clock and thermostat.

- Households appeared to vary in the amount of use made of the gas fire. Some used it regularly, either in the morning or in the evening, while others reported that they made little use of it.

- Room thermostat settings were very high in two houses (25°C and 30°C).

- All of the water cylinder thermostats that were examined were found to be set at 60°C or above and several occupants observed that this was, if anything, too hot. Even those who had a good understanding of their system were not aware of the role of the water thermostat and many did not know that one was fitted. No households had attempted to adjust it since the system was installed, despite concerns about high water temperatures.

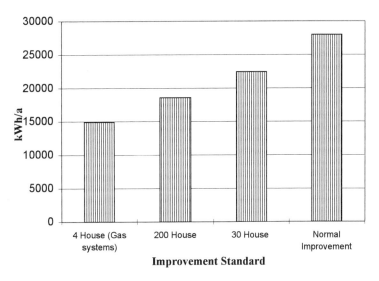

Figure 8.7 **Comparison of total energy consumption for all schemes**
Source: Bell and Lowe 1995

127

The three schemes compared

Figures 8.7 and 8.8 compare temperatures and energy consumption for each of the schemes against the base line set by the modernisation standard which was used in York at the beginning of the project. This comparison shows that in the type of low rise, traditionally constructed houses featured in this project, the application of well established methods of improving energy efficiency can reduce energy consumption by between 40 per cent and 50 per cent when compared with a modernisation standard which is typical of many local authority schemes. The additional costs of improving modernisation standards in this way are not excessive, with simple pay back periods of less than 10 years in most cases. With levels of insulation similar to those achieved in the 4 house scheme, particularly where the house is small, a full central heating system may not be necessary giving a substantial saving on capital cost (see discussion of Gas House B - unit heater system - above).

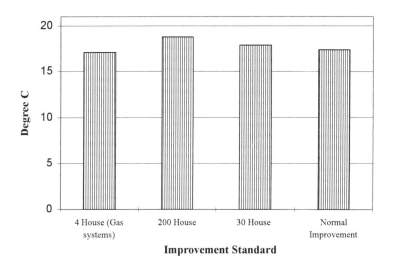

Figure 8.8   Comparison of temperatures in all schemes
Source: Bell and Lowe 1995

The comparison of temperatures presents a very even picture across all standards and indicates an obvious desire for comfort whatever the level of insulation. This means, that in broad terms, the efficiency improvements at York resulted in real cash savings to tenants in the houses with additional energy works.

The York project in context: a conclusion

Figure 8.9 shows how the best of the houses at York compare with energy use in the average dwelling (GB data for 1991), and in two other low energy housing projects. The Pennyland scheme represents one of the best low energy housing projects constructed in the UK during the 1980s, while Kranichstein, (Darmstadt, Germany), represents the best of the low energy housing projects currently being undertaken in Europe. Pennyland addressed the concerns of the 1970's, which were the exhaustion of fossil fuels, and security of supply. Kranichstein addresses the much more demanding agenda of the 1990's, which is the stabilisation of atmospheric carbon dioxide concentration. York outperforms the GB average, and approaches the energy performance of the Pennyland project. It does, however, fall a long way short of the target set by the houses built at Kranichstein.

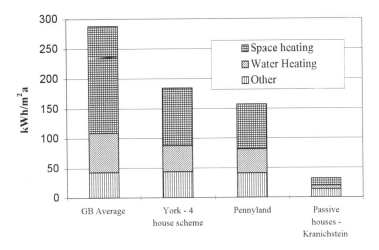

Figure 8.9   Comparison of energy standards
Source: Bell and Lowe 1995[3]

The agenda at York was to implement an energy efficiency programme at modest cost, that could be undertaken by the Local Authority acting alone, within the constraints imposed by existing housing modernisation programmes. The York Project concentrated on space heating and to a lesser extent water heating, using the technology of the 1980's. While there is still considerable scope for improvement in window and door performance, the York Project has taken this approach almost as far as it will go. The tools for improving the energy efficiency of the existing housing stock are now available in abundance and if they are used effectively will bring large benefits for the environment as well as providing affordable warmth for many people. If the projections of climate change discussed in chapter 2 are correct however, it will be necessary to go even further than we were able to at York. Achieving levels of energy efficiency similar to those at Kranichstein is perhaps the next challenge for existing as well as new housing.

Notes

1   Since all the houses in the 30 house scheme are heated by gas, only the gas houses from the 4 house scheme have been included in this analysis. In addition, the comparison being made is between the 4 house level of energy efficiency and the level of efficiency which is typical of York's normal modernisation standard. This is not the same as a before-and-after comparison because normal modernisation involves heating system efficiency and fuel switching improvements. This is arguably a more useful comparison, in that it allows the costs and benefits of a change in modernisation policy to be more readily defined.

2   Electricity use in the 2 gas heated houses in the 4 house scheme was roughly half that in the 30 house scheme. Since electricity consumption in these houses was mainly for lighting and appliances, the efficiency measures installed are unlikely to have affected electricity consumption in any significant way. The cost and carbon dioxide comparisons are therefore based on differences in gas consumption only

3   The sources for the individual data items in this table are as follows:
   • GB average was compiled for 1991 from Shorrock and Brown (1993) and DoE (1993a).
   • Pennyland data is taken from Lowe et al. (1985)
   • Data on the Kranichstein passive houses are from Feist (1994)

9 Case study 2: The Longwood Low Energy House

Robert Lowe & Steve Curwell

Introduction

The Longwood House is a well insulated, air-tight, two storey detached house, in a simple vernacular style in the North of England. Constructed in load bearing masonry, it is one of the most energy efficient houses to have been built in the UK to date. This chapter reviews the process that lead to the construction of the house, describes its design and construction, and presents information on its energy use.

Background

The Longwood project arose as a result of the desire of builders Steve Slator and Bill Butcher to establish a market lead for energy efficient and environmentally friendly house construction in the East Pennine area. Planning permission had been acquired in 1991 for a four bedroom house of approximately 107 m² gross floor area, on an almost perfect passive solar site, steeply sloping toward the south.

The environmental objective of the design was to minimise the environmental impact arising from energy use and from choice of materials. This objective had to be pursued subject to a number of constraints. The most important of these were:

- The house was to be built on a speculative basis. In the judgement of the builders, marketing considerations required that the house not look out of the ordinary. They specifically wanted to avoid an overtly low-energy appearance, which they considered unsaleable on this site.

- The speculative nature of the project, and minimal external financial support meant that it was important to choose the most cost effective package of energy saving measures. It was in particular not feasible to consider active solar measures, and construction had to be based on traditional methods with which the builders were familiar.

- The steep slope of the site resulted in a dwelling partially cut into the hillside, with the entrance at first floor on the north facade. The cost of a retaining wall was an additional financial constraint on the project.

The urban context of the site was dominated by traditional weavers' cottages and other vernacular stone buildings, together with more recent speculative housing in reconstituted stone. The former, with their rows of stone mullioned windows, shallow plans, and small eaves overhangs provided much of the inspiration for the Longwood house. Figure 9.1 shows the house and its immediate surroundings, from the south-west.

Figure 9.1   The Longwood House (south view)

Detailed design

In this context, load bearing masonry construction appeared to be the most appropriate choice on technical and aesthetic grounds. The design constraints outlined above placed a fairly severe limit on the amount of glazing that could be place on the south facade of the house. The house as-built has approximately 20 per cent glazing on this facade.

The experience of the authors over the previous decade included both the Pennyland project in Milton Keynes (Lowe et al 1985) and the Salford low energy houses (Webster 1987). Low space heat demand was achieved in these schemes by building in high insulation levels, avoiding cold bridging wherever possible, detailing for air-tightness, and in the case of Pennyland, by using the most efficient fossil fuelled heating system available. The concern of Butcher and Slator to achieve high performance using traditional construction methods was entirely compatible with this approach. Work at Pennyland and Great Linford (Everett et al. 1985) showed that with double glazing of modest thermal performance, the space heating consumption of a well-designed house was almost independent of total glazing area.

Aspects of the design of the house which related to air-tightness were informed by the authors' research in this area (Lowe et al 1994). In 1991, preliminary results from this project indicated that wet construction offers potentially very high levels of air-tightness. This contradicts the still widely held belief that air-tightness is best achieved using Scandinavian and North American timber framed construction methods, and it is therefore worth saying a few words to back the bald assertion.

The simplest way of measuring air-tightness in buildings is by pressurisation testing (see for example. Steven 1989). This involves the use of a fan to establish a pressure difference across the thermal envelope of the building under test, and then measuring the air flow required to maintain that pressure difference. Standard leakage rates are normally quoted for a test pressure difference of 50 Pascal (Pa), in units of air changes per hour (ac/h). Leakage rates for English houses built since the 1970s are in the region of 10-20 ac/h, with a mean of around 14 ac/h (Perera and Parkins 1992). It is rare to achieve 50 Pa leakage rates as low as 7 ac/h, which is the level adopted by the Electricity Industry for its Medallion Home standard. This is, in turn, some way above the 1980 Swedish Building regulation requirement for single dwellings of 3 ac/h. However a simple plastered wall is so air-tight that laboratory measurements of air leakage are almost impossible to carry out. If one were to build a dwelling out of nothing but wet-plastered wall, the 50 Pa leakage rate would be in the region of 0.1 ac/h (Lowe et al. 1994). The reason why masonry construction so often fails to perform, is to do with leakage at junctions, and through those parts of the dwelling that are timber framed. Most masonry dwellings are composite structures, with timber floors and roofs, and of course windows and doors. Air-tightness therefore requires careful design and construction of these details. Provided this is done, it is, in the view of the authors, probably easier to achieve air-tightness in masonry than in timber framed construction.

Work for the Longwood House therefore concentrated on achieving low U values, detailing to limit air leakage and avoid cold bridging. The choice for the wall construction was a 150mm fully-filled wall cavity, using nylon wall ties. These were supplied by K.G. Kristiansen, Kolding, Denmark. To the authors' knowledge no equivalent product for this size of cavity is available in the UK. Insulation in the wall cavities was carried past the level of the damp-proof membrane down to the footings. This minimised the cold bridge that would otherwise have occurred at the junction between the ground floor and external wall. Negotiations with the Building Inspector were straightforward, and there were no problems gaining approval for this construction. The roof was to be insulated at ceiling level with 300 mm of blown cellulose. The ground floor slab was cast on 100 mm expanded polystyrene insulation. The windows were timber, supplied by a local firm of joiners, with 20 mm low emissivity double glazed sealed units. The inner leaf of the wall and all internal partitions were constructed of 100mm dense clinker block to maximise internal thermal mass and reduce sound transmission between rooms. Clinker blocks also have a significantly lower embodied energy content than lightweight expanded concrete blocks.

Table 9.1
Insulation levels at the Longwood House

elemental U values	W/m^2K
wall	0.21
roof	0.11
floor	0.23
glazing	2.4

Figure 9.2 shows a foundation-to-eaves construction of the house. The nominal U values for the major elements of the fabric of the house are shown in Table 9.1. The avoidance of cold bridging means that actual U values should be close to those shown.

Achieving air-tightness

The possibilities of an in-situ cast or beam-and-block concrete first floor were discussed. Both options, but particularly the first, promised higher levels of air-tightness than the timber alternative. Eventually timber was chosen because of lower cost and the greater familiarity of the builders with this method. Timber floors are likely to be less air-tight than the concrete alternatives, because the external wall in the joist space is not plastered, and is therefore 100-1000 times more leaky than the plastered wall in the rooms above and below. In order to counteract this effect, the authors suggested that the edge of the first floor be sealed with in-situ foamed polyurethane foam. This was achieved by fixing noggins between the joist ends approximately 25 mm from the external wall, and foaming the space between the noggins and the wall from the underside before the fixing of the ground floor ceiling. A better seal may have been achieved if the foam had been applied after the fixing of the ceiling, through holes drilled in the floor above, though laboratory tests showed that this approach was more difficult to control.

A number of steps were taken to achieve an air-tight first floor ceiling. The most important of these was that the ceiling was fixed before the first floor internal partitions were completed. This eliminated leakage from junctions between ceiling plasterboard and partitions. The domestic hot water system is unvented, reducing the number of service penetrations through the ceiling. The loft hatch is a pre-fabricated draught proofed unit. Considerable care was taken by the builders to seal the remaining penetrations through this element of the house - in particular plywood boxes were fitted around the two soil stacks as they passed through the ceiling, and these were then filled with polyurethane foam.

Air-tightness at window reveals was achieved by careful construction. The window reveals were thoroughly thermally broken - the inner leaf was not returned, and separate lintels were used for inner and outer leaves of the wall. Cavity closure at the jambs and head was achieved with plasterboard, which was bedded on a continuous bed of bonding plaster at the inner leaf, and into a continuous bead of mastic gunned into a channel in the timber window frame. The internal wooden window sill was treated in a similar fashion. Here again, the authors originally suggested an alternative detail, pioneered in Germany, in which the wall cavity is closed with a timber carcase, into which the window frame proper is subsequently set (Arbeitsgemeinschaft für zeitgemässes Bauen 1989). The builders preferred their own rather more conventional solution, which appears to have worked very well.

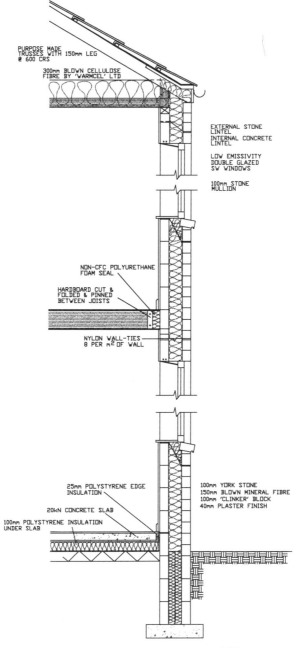

PURPOSE MADE
TRUSSES WITH 150mm LEG
@ 600 CRS

300mm BLOWN CELLULOSE
FIBRE BY 'WARMCEL' LTD

EXTERNAL STONE
LINTEL
INTERNAL CONCRETE
LINTEL

LOW EMISSIVITY
DOUBLE GLAZED
SW WINDOWS

100mm STONE
MULLION

NON-CFC POLYURETHANE
FOAM SEAL

HARDBOARD CUT &
FOLDED & PINNED
BETWEEN JOISTS

NYLON WALL-TIES
8 PER m² OF WALL

25mm POLYSTYRENE EDGE
INSULATION

20kN CONCRETE SLAB

100mm POLYSTYRENE INSULATION
UNDER SLAB

100mm YORK STONE
150mm BLOWN MINERAL FIBRE
100mm 'CLINKER' BLOCK
40mm PLASTER FINISH

Figure 9.2   Vertical section through the Longwood House

Choice of materials

Considerations of environmental impact affected choice of materials in a number of areas. A decision was taken to minimise the use of fibreboard and chipboard in the house, in an attempt to reduce the number of sources of formaldehyde gas. This particularly affected the construction of the floors, confirming the choice of concrete for the ground floor, and resulting in a first floor of tongued-and-grooved timber rather than chipboard. Timber windows were used instead of aluminium or PVC, resulting in reduced embodied energy, and elimination of a number of specific classes of emissions from the production process (vinyl chloride monomer in the case of PVC, fluorides, halogenated carbon compounds, and mining waste in the case of aluminium production). Timber was finished internally with water based stains or solvent-free paints. Treatment of external timber remained a problem, and conventional microporous stains were used. Wall insulation was blown mineral fibre, which should be inert and, encased in the building fabric, harmless in use. Roof insulation was blown recycled newsprint supplied by Warmcel.

Heating and ventilation

With a building designed for air-tightness, it is necessary to make positive provision for ventilation. Two alternatives were discussed - whole house mechanical ventilation with heat recovery (MVHR), and passive stack ventilation with local mechanical extract and trickle vents. In a gas heated house with currently available MVHR systems, and with no certainty of achieving a 50 Pa leakage rate of 3 ac/h or less, primary energy use was not predicted to be strongly affected whichever option was chosen. Though a full MVHR system was offered by the builders as a marketing option, in the end the passive stack system was installed due to its greater simplicity and lower first cost. The system was installed according to the recommendations of BRE Digest IP 21/89, since superseded by IP 13/94.

The heating system chosen for the building was a gas fired wet central heating system with time clock, room thermostat and thermostatic radiator valves. In addition to the gas fired central heating system, the owners of the house installed a small solid-fuel-effect gas stove in the living room. The option of installing a condensing rather than a conventional gas boiler was considered carefully. At the time the additional capital cost of the former was approximately £400, and the marginal pay back time considered against a good fan-flued low thermal capacity non-condensing boiler was predicted to be in the region of 7-15 years. The decision was finally tipped in favour of the condensing boiler by the offer of a favourable discount by the manufacturer, Stelrad. Shortly after the boiler was installed, the Energy Saving Trust introduced a £200 subsidy for installers of condensing boilers. The house was BREEAM rated by BRE, and was awarded a score of 9.3 on the NHER scale.

Construction

The builders reported that construction was straightforward. Having wall cavities post-filled minimised the amount of insulation material on site during construction, and eliminated risks from weather on a very exposed site. There were no problems with the wide wall cavity, and inspection of the cavity after construction showed it to be very clean. This appears to be a general phenomenon with wide wall cavities. The plastic wall ties were cheaper to buy than stainless steel, and safer and more pleasant to use, though there were more of them than one would find in a standard 50 mm cavity. Construction problems that did occur resulted from the retaining wall required by the steep slope of the site.

Evaluation

The house was completed in spring of 1992, but not occupied until early summer 1993. Leeds Metropolitan University were able to pressure test the house shortly before occupation. The test was undertaken on the 8th April 1993 using the University's Minneapolis blower door, under near perfect weather conditions. The 50 Pa leakage was estimated as 2.95 ± 0.3 ac/h. This was achieved blind, without the need for remedial work, making the Longwood House one of the most air-tight to have been constructed in the UK. More air-tight construction has been documented on three occasions: in pre-fabricated panel timber framed houses at Two Mile Ash (Ruyssevelt et al 1987), in a single storey timber framed house on Orkney (Scivyer et al 1994), and, following remedial work, in load bearing masonry at Charlbury in Oxfordshire (Olivier 1994). It is possible that a number of other houses would prove to be as air-tight, but fan pressurisation is still not routine in the UK.

Air leakage was traced using a smoke pencil during the pressurisation testing. Leakage was predominantly around the edges of the two floors. Leakage around the edge of the ground floor slab was somewhat unexpected. It is likely to have been exacerbated by the insertion of a thin layer of expanded polystyrene between the floor slab and the inner leaf of the wall, which was intended to minimise cold bridging. In future dwellings, the authors would recommend either sealing this junction, or omitting this detail completely and casting the floor slab directly against the inner leaf of the external wall. There was very little leakage around the windows. Leakage was noticeable around ceiling roses (these were not caulked), and from electrical sockets. The latter is exacerbated by the use of open electrical conduit. At a test pressure of 150 Pa, leakage was noticed from behind the wall-string of the staircase. This ran up the external wall on the north side of the house, and was another area not plastered.

If these residual leaks were caulked, it is likely that the leakage rate of the Longwood House could be reduced substantially - but with passive stack ventilation there is little point in going further. The 50 Pa leakage rate with the trickle vents open was approximately twice as high as the 3 ac/h measured with all vents closed. This suggests that operation of the trickle vents can exert considerable control over the ventilation rate of the house.

Table 9.2
Comparison of energy performance and targets

	electricity (kWh)	gas (kWh)	total delivered (kWh)	delivered energy index (kWh/m²)	carbon dioxide emission (kg/m²/a)
GB mean (1991)	(not applicable)		23325	288	87
Pennyland 2	2600	11530	14130	157	44
BRE mandatory standard	(not applicable)				29-31
Longwood	2239	6355	8594	80	25
BRE optional standard	(not applicable)				17-21

Energy use in the house has been determined from energy bills provided by the owner. Based on these, energy use for the period September 1993 to September 1994 has been calculated. The figures are presented in Table 9.2, together with GB mean data (taken from Shorrock and Brown 1993), and data from the Pennyland field trial (Lowe et al 1985). The latter is an interesting comparison since it represents a level of thermal insulation and air-tightness slightly beyond that required by current Building Regulations (100mm fully filled wall cavities, 150 mm loft insulation, double glazing and an air leakage of 5 ac/h at 50 Pa). The figures from this field trial are for 3-bedroom houses, with a gross floor area of approximately 90 m². The average floor area for all dwellings is approximately 81 m² (DOE 1993a). Finally, we compare these measured energy consumptions with the requirements of the Building Research Establishment' s Environmental Standard for housing (Prior and Bartlett 1995).

Delivered energy use in the Longwood House is approximately 39 per cent of the GB average, and about 61 per cent of the average for Pennyland Area 2 houses. The delivered energy index (delivered energy per m² of gross floor area) for the house is about 29 per cent of the GB average, and about 51 per cent of that of an average Pennyland Area 2 house. Gas use is reduced by about 45 per cent in absolute terms compared with an average Pennyland Area 2 house. Although room temperatures at Longwood have not been measured continuously, spot measurements suggest that they are at least as high as those maintained at Pennyland.

Carbon dioxide emissions from the Longwood House are just over 25 kg(carbon dioxide)/m²/a. The house therefore exceeds the Building Research Establishment's Environmental Standard Award mandatory target of 29 kg(carbon dioxide)/m²/a, but does not achieve the optional standard for a house of this size of 17 kg(carbon dioxide)/m²/a (Prior and Bartlett 1995).

The evaluation shows that the primary objectives of the project were achieved. The builders have successfully constructed a low energy house of very good performance, at small additional cost. They have been able to sell this on the open market at a modest profit in the second deepest economic recession of this century. The present owners of the house found the anticipated energy savings and low environmental impact an incentive to buy the house, and have taken an interest throughout in its energy performance.

Lessons for the future

A number of lessons can be learnt from the Longwood House. Simple strategies worked extremely well in terms of reducing energy use. High levels of thermal insulation were incorporated into a load bearing masonry construction with few assembly difficulties, and no special techniques or equipment. Furthermore, a committed and technically competent builder were able to construct a load-bearing masonry dwelling with a 50 Pa air leakage rate of around 3 ac/h, in a commercial situation.

The Longwood House represents a good starting point for low energy house design, and as a model for house builders provides an achievable standard of energy and air-tightness performance for loadbearing masonry construction. Further reductions in energy use can however be achieved in all areas. Reductions in space heating requirement equivalent to a little under 1000 kWh/a of gas could be achieved by improved windows and doors (modern high performance windows have U values less than half the 2.4 W/m²K achieved at Longwood). Minor reductions in water heating could be achieved by the use of aerating taps and a better insulated storage cylinder, but large reductions in water heating would require the use of solar water heating. However, major overall reductions in carbon dioxide emissions would require a much more thorough-going approach to reducing the electricity used by lights and household appliances, which were responsible for 56 per cent of emissions from the Longwood House.

10 Conclusions

This book has attempted to look in theory and in practice at the nature of the energy efficiency challenges which face the housing sector and those who work within it. These are on the one hand the problem of climate change, and on the other the ill-health and discomfort suffered by occupants of substandard, energy inefficient housing. At the time of writing, the world scientific community meeting in Madrid has agreed, for the first time, that anthropogenic climate change is occurring. Our analysis of this problem suggests that very large reductions in carbon dioxide emissions from industrialised countries will be needed if the composition of the global atmosphere is to be stabilised. Housing, which is directly responsible for almost one third of emissions, will need to play a full part in achieving these reductions.

The scale and seriousness of the problem of fuel poverty in the UK are indicated by the following facts. It appears that only 13 per cent of households are able to maintain an average temperature of 21°C in living rooms at an outside temperature of 2°C. Boardman estimates that 30 per cent of households spend more than 10 per cent of their income on energy, and are therefore likely to suffer from a combination of low winter temperatures, and damp, mouldy and unhealthy housing. The evidence suggests that the very young, the elderly and the infirm are most at risk from these conditions. Finally, although there is no simple causal link between cold housing and mortality, excess winter deaths in the UK have remained between twenty and sixty thousand over the last 20 years.

At first sight, the goals of achieving climate stabilisation and the elimination of fuel poverty, appear to be irreconcilable. The former requires a 75 per cent to 90 per cent reduction in carbon emissions from energy use, while the latter appears to require increased energy use, among a large section of the UK population. Much of this book is devoted to a description of technologies which in the long term will enable these goals to be achieved simultaneously. We show how all areas of domestic energy use, space heating, water heating, lighting and energy use by electrical appliance can be reduced, and we quote one example which demonstrates the achievement of almost 90 per cent reductions in delivered energy use and carbon emissions in new housing, by the ruthless application of best available technology. We also show how application of a very modest series of measures in existing housing can halve energy use and carbon dioxide emissions.

Meeting this joint challenge will require the development of housing policy across a broad front, in order to achieve a shift in investment to the housing sector. We have argued that investment in housing energy efficiency which brings affordable warmth could have the effect of reducing resource requirements elsewhere, notably in the health sector, and of generating employment benefits. This same investment can bring reductions in expenditure on the repair and maintenance of dwellings as energy efficiency reduces the wear and tear on the building fabric. The resources released will be long term and will also bring choices, some of which will be hard ones to make. For example, the resources released in the health sector could be put back into the health sector in the form of more advanced treatments and care systems, or they could be used to develop the energy efficiency of the housing stock which would, further reduce the load on the health system.

While recognising the difficulties involved in accounting explicitly for these indirect effects, we argue that attempting to identify and quantify them will assist the process of building a political consensus in favour of energy efficiency in housing.

Not all issues are resource issues. Social and moral concerns exist over the ways in which energy resources are used and distributed. These issues will be crucial in ensuring that any transition to a low carbon dioxide economy is achieved without significant social unrest or breakdown. We find it hard to envisage a satisfactory solution to the problem of climate change which does not address associated problems of equity and resource distribution.

Understanding the emerging concern with the development of sustainability in the future development of communities will also have important implications for housing. The ways in which communities are designed and evolve, and the relationships between housing, transport, employment and other systems will play an important part in the overall reduction of carbon dioxide . To take a relatively simple example, the development of combined heat and power systems will influence housing development and will raise questions of ownership and control of the provision of heat and power.

The way forward

One of the lessons of the last twenty years is the way in which technological developments have continually pushed back the limits to energy efficiency. Although any attempt to predict the future path of technological development is likely to be proved at least partially wrong, we believe that we can identify a number of developments that offer great promise in the short term. These include:

- advanced glazing systems, with whole window U values of 1.0 W/m²K or less, which offer the possibility of significantly improving the energy performance of existing and new dwellings at modest additional cost;

- improved air-tightness in houses coupled with the controlled ventilation systems (both passive and mechanical), to ensure good air quality and humidity control with minimum energy use;

- high efficiency electrical appliances and lighting, which can significantly reduce the electricity requirement of new and existing housing;

- energy supply technologies such as combined heat and power, which can significantly reduce the environmental impact of energy supply; the application of combined heat and power in large 60s and 70s housing blocks is developing with encouragement from the UK government, but the technology is yet to be exploited to any significant degree in low energy housing.

Future developments, for example, of cheap active solar domestic hot water systems, using transparent insulation materials, and with the integration of photovoltaics offer exciting additional possibilities in the future. However, the most important component of energy efficient housing is, and in our view will remain, a well insulated and air tight thermal envelope. Although the techniques for achieving this in new and existing housing are well understood in principle, achieving it in practice is often problematic. Experience shows that good intentions are frequently compromised by poor workmanship and inadequate detailed design. Adopting a strategy of energy efficiency improvement without giving consideration to the education and training needs of the people concerned at all levels in the industry, is unlikely to be effective.

The problem is one of how to transfer the available technology from the research and development community to practice. This process is proving to be disappointingly slow. The reasons for this lie not in the technology as narrowly understood, but in the realms of individual and social behaviour, institutional systems, and energy and environmental policy.

It is, of course, people who use energy and who decide what, if any, improvements they make to their houses and other buildings. Despite a consistent message that energy efficiency investment brings greater comfort and reduces energy consumption (often with short pay back periods), why is the take up of investment (particularly within the existing housing stock) so low? What are the motivational factors involved and why do they seem to be lacking? The literature provides some insight into these questions. It seems that the problems are as much to do with 'hassle', uncertainty and inertia factors as with finance.

As discussed in chapter 4, there is significant uncertainty about the part played by energy efficiency in decisions made by house buyers. This also applies to the purchase of domestic appliances. Energy labelling, whether applied to houses or the appliances used within them, is a start, but unlikely on its own to change radically the way producers and suppliers view energy efficiency when making supply decisions. Making an energy efficient house more marketable would add to the incentive for all owners of housing to invest.

A clear message from behavioural research is that the provision of advice and guidance coupled with systems which are understandable to all members of a household can help to reduce energy consumption. However questions arise such as: What is the best method of advice delivery? How can the effects of advice be sustained in the long term? What effects on life-style are likely to result and are these considered reasonable and equitable? What would be the behavioural variation in houses with high levels of insulation? Central to these concerns is the relationship between hardware and people. For example, detailed ergonomic work on the design of heating controls or the display of energy consumption information would be of benefit in helping people make more effective use of energy in their homes. At all levels, it is necessary to understand the behavioural dimension if we are to make the most of the investments made in energy efficient technology.

Facing the future

The need for energy efficiency in the domestic sector is becoming harder to doubt. Even without the environmental imperative, concerns about health constitute sufficient grounds for pursuit of this goal. As a result, energy efficiency is to a large extent a 'no-regrets' option.

It is in our view pointless blaming individuals for not buying and using the most energy efficient technology, if it is not available on the market, if the intermediaries (architects, planners, surveyors, engineers and shop assistants) have not heard of it, if it is uneconomic to install against a background of static or falling energy prices, and if the individuals in question do not have sufficient discretionary income to afford it in the first place. The government's attempts to encourage individual responsibility for energy conservation, exemplified by the slogan 'Helping the Earth begins at home', may be at best cosmetic in the light of these difficulties. We believe that the way forward is clear, and that the road begins with government, at all levels, and with the political process.

References

ALDERSON, M.R. (1985) Season and Mortality. *Health Trends,* vol. 17 pp. 210-24.

ALPER, B. FORREST, R.A. and LONG G. (1985) Active Solar Heating in the UK: Department of Energy R&D Programme 1977-1984. ETSU R25.

AMA (1985) An Energy Policy for Housing. London, Association of Metropolitan Authorities.

AMA (1991) Housing, Energy and the Environment: A Strategy for Action. London, Association of Metropolitan Authorities.

ANDERSON B R, (1988) British Standard Code of Practice for Energy Efficient Refurbishment of Housing. Energy Efficiency in Buildings, Information Leaflet No 7, 1988 (updated 1991). Building Research Establishment, Watford, UK

ANDERSON B.R. (1985) Energy Assessment for Dwellings Using BREDEM Worksheets. BRE IP 13/85, Building Research Establishment, Watford, UK.

ANDERSON, T.W. and LeRICHIE, W.H. (1970) Cold weather and myocardial infarction. *Lancet,* February 7th 1970, p 291.

ANON (1993) Saskatchewan Advanced House. Saskatoon, Canada, Sun Ridge Group.

ANON (1994) Hockerton Housing Scheme Goes to Ground. *Building Services,* vol 16 (7).

ARBEITSGEMEINSCHAFT FUER ZEITGEMAESSES BAUEN (1989) Ökologisches Bauen: Umweltverträgliche Baustoffe. Arbeitsgemeinschaft für zeitgemässes Bauen, 2, 183.

ASHLEY S. (1988) What the doctor ordered. *Building Services,* vol 10 (4).

ATKINS W.S. & PARTNERS (1984) Combined Heat and Power District Heating Feasibility Programme: Stage 1. Summary Report and Recommendations. Department of Energy, Energy Paper 53, London, HMSO.

BACH, W. (1980) The CO_2 Issue: What are the Realistic Options ? *Climate Change*, 3, 3-5.

BAKER, N. and STANDEVEN, M. (1995) A Behavioural Approach to Thermal Comfort Assessment in Naturally Ventilated Buildings. *In* Proceedings of CIBSE National Conference, Eastbourne, 1995. Charted Institution of Building Services Engineers, London.

BALCOMBE J.D. (1980 & 1982) Passive Solar Design Handbook. 3 volumes. National Technical Information Service, Springfield VA, USA.

BECKER, L.J. SELIGMAN, C. FAZIO, R.H. and DARLEY, J.M. (1980) Relationship Between Attitude and Residential Energy Use. *Environment and Behaviour,* vol 10 (cited in Olsen, 1981).

BEDFORDSHIRE COUNTY COUNCIL (1994) Structure Plan 2011: Environmental Appraisal. Bedford, Bedfordshire County Council.

BELL, M. and LOWE, R.J. (1993) Conserving Energy in the Existing Housing Stock: The York Energy Demonstration Project. *In* A Focus for Building Surveying Research. Proceedings of a seminar held at the Royal Institution of Chartered Surveyors, London, RICS.

BELL, M. and LOWE, R.J. (1995) The York Energy Demonstration Project: Final Report to York City Council. Leeds Metropolitan University, Faculty of Design and the Built Environment, Leeds, UK.

BELL, M. LOWE, R.J. and ROBERTS, P. (1992) Conserving Energy in Local Authority Housing: The York Energy Demonstration Project. *In* Housing Technology and Socio-Economic Change. edited by A. Middleton and O. Ural, Proceedings of the 20th. Working Congress on Housing, Birmingham, University of Central England.

BERKOWITZ, P.I. KARL, S.L. and RAMSAY, J. (1994) Capturing Conservation Through Community Energy Management. *Home Energy,* March/April 1994.

BERTHOUD, R. (1983) Disconnection. *In* Energy and Social Policy. edited by J. Bradshaw and T. Harris, London, Routledge & Kegan Paul, pp 69-84.

BLACK, J.S. STERN, P.C. and ELWORTH, J.T. (1985) Personal and Contextual Influences on Household Energy Adaptations. *Journal of Applied Psychology,* vol 70 (1), pp 3-21.

BLOWERS, A. (ed) (1993) Planning for a Sustainable Environment. London, Earthscan.

BLUMSTEIN, C. KRIEG, B. SCHIPPER, L. and YORK, C. (1980) Overcoming Social and Institutional Barriers to Energy Conservation. *Energy,* vol 5, pp 355-371.

BLUNDEN, J. AND REDDISH, A. (eds) (1991) Energy, Resources and Environment. London, Hodder and Stoughton.

BOARDMAN, B. (1991a) Fuel Poverty. London, Belhaven Press.

BOARDMAN, B. (1991b) Lessons from Ten Years Cold: A Decade of Fuel Poverty. Newcastle upon Tyne, Neighbourhood Energy Action.

BOARDMAN, B. (1993) Prospects for Affordable Warmth. *In* Unhealthy Housing: Research, remedies and reform. edited by R. Burridge and D. Ormandy. London, E&FN Spon, pp 382-400.

BOLIN, B. et al (1986) The Greenhouse Effect, Climate Change and Ecosystems. Scientific Committee on Problems of the Environment SCOPE 29, Wiley, Chichester, UK.

BOSS, A. ORBASLI, A. and SUTCLIFFE, S. (1993) House Design Studies Overview. Department of Trade & Industry, ETSU S 1362.

BOSSELAAR L. (1994) The Role of the Government. *Sun at Work in Europe*, vol 9 (1).

BRADSHAW, J. and HUTTON, S. (1983) Social Policy Options and Fuel Poverty. *Journal of Economic Psychology*, vol 3, pp 249-266.

BRE (1975) Energy conservation: a study of energy consumption in buildings and possible means of saving energy in housing. BRE CP 56/75, Building Research Establishment, Watford, UK.

BRE (1990) Climate and Site Development. BRE Digest no. 350. Building Research Establishment, Watford, UK.

BRE (1993) BRE Housing Design Handbook: Energy and Internal Layout. BRE 253, for Energy Efficiency Office, Best Practice Programme.

BRE (1994) Continuous mechanical ventilation in dwellings: design, installation and operation. BRE Digest no. 398. Building Research Establishment, Watford, UK.

BRECSU (1990) Cavity Wall Insulation in Existing Dwellings. Good Practice Case Study 3, London, Department of the Environment.

BRECSU (1992) Affordable New Low Energy Housing for Housing Associations. Future Practice R&D no.2, Best Practice Programme, Energy Efficiency Office and National Federation of Housing Associations, London, HMSO.

BRECSU (1993a) Benefits to the Landlord of Energy Efficient Housing: Merseyside Improved Homes. Good Practice Case Study 155, London, Department of the Environment.

BRECSU (1993b) Benefits to the Landlord of Energy Efficient Housing: Waltham Forest. Good Practice Case Study 186, London, Department of the Environment.

BRECSU (1993c) Benefits to the Landlord of Energy Efficient Housing: Glasgow. Good Practice Case Study 187, London, Department of the Environment.

BRECSU (1993d) Benefits to the Landlord of Energy Efficient Housing: Bristol City Council. Good Practice Case Study 189, London, Department of the Environment.

BRECSU - OPET (1992) Energy Efficient Lighting of Buildings. BRECSU on behalf of Commission of the European Communities DG XVII, Watford, UK.

BREHENY, M. (1992) The Compact City: An Introduction. *Built Environment,* vol 18 (4).

BREHENY, M. (1993) Planning the Sustainable City Region. *Town and Country Planning,* vol 62 (4).

BREHENY, M. and ROOKWOOD, R. (1993) Planning the Sustainable City Region. *In* Planning for a Sustainable Environment. edited by A. Blowers, London, Earthscan.

BROOKE, J. (1991) Policy and Persuasion: What a Green Housing Policy Might be and How to Persuade Policy Makers to Adopt It. Glasgow, City Housing Department.

BROWN, H. (1993) Evaluation of Passive Solar Potential Multi-residential Dwellings and Domestic Retrofit Measures. ETSU S1352.

BRUNEFREEF, R. DOCKERTY, D.W. SPEIZER, F.E. WARE, J.H. SPENGLER, J.D. and BENJAMIN, G.F. (1989) Home Dampness and Respiratory Morbidity in Children. *American Review of Respiratory Diseases,* vol 140 (5), pp 1363-7.

BRUNNER C.U. & NÄNNI J. (1990) Wärmetechnische Massnahmen zur Optimierung der Gebäudehülle. *in* Energiefforschung im Hochbau. Schweizerisches Status-Seminar 6. EMPA-KWH. September 1990.

BS 5449 (1990) Forced Circulation Hot Water Central Heating Systems for Domestic Premises, British Standards Institution, Milton Keynes, UK.

BUILDING (1995) Construction's Priorities for 1995. *Building Awards Special Supplement, Building,* 21 April 1995. The Builder Group, London.

BULL, G.M. and MORTON, J. (1978) Environment, Temperature and Death Rates. *Age and Ageing,* vol 7 pp. 210-224.

BUNN R. (1994) Living on Auto. *Building Services,* vol 16 (7).

CADDET (1994) Utilities help home-owners to cut energy consumption by 12%. Sittard, The Netherlands, Centre for the Analysis and Dissemination of Demonstrated Energy Technologies, Result 174.

CAMPBELL, R. (1993) Fuel Poverty and Government Response. *Social Policy & Administration,* vol 27 (1), pp 58-70.

CARLSSON, B. ELMROTH, A. and ENGVALL, P.A. (1980) Airtightness and Thermal Insulation: Building Design Solutions. Stockholm, Statens Råd för Byggnadsforskning.

CARPENTER, P. (1988) Caer Llan Berm House. Peter Carpenter, Caer Llan Field Studies Centre, Near Monmouth, UK.

CARPENTER, S. and KOKKO, J. (1993) Design of the Waterloo Region Green Home. *In* Proceedings Innovative Housing Conference, Vancouver, Canada.

CHAPMAN, P.F. (1990) The Milton Keynes Energy Cost Index. *Energy and Buildings*, vol 14 (2), pp 83-101.

CHELL, M. and HUTCHINSON, D. (1993) London Energy Study: Energy use and the Environment, Prepared under the Commission of the European Communities' Urban and Regional Energy Management Programme. London, London Research Centre.

CHRISTENSEN B. & NORGÅRD J. (1976) Social Values and the Limits to Growth. *Technological Forecasting and Social Change*, vol 9, p411.

CHRISTENSEN, B. and JENSEN-BUTLER, C. (1980) Energy, Planning of Heating Systems and Urban Structure. London, Regional Studies Association.

CHRISTIE, I. and RITCHIE, N. (1992) Energy Efficiency: The policy agenda for the 1990s. London, Policy Studies Institute.

CLARKE, J.A. (1985) Energy Simulation in Building Design. Adam Hilger, Bristol and Boston.

COLE, J. (1993) Canada's R-2000 Program: What is it? What has it accomplished? What could it do for You? *In* Proceedings Innovative Housing Conference, Vancouver, Canada.

COLLINS, K.J. (1993) Cold- and Heat-Related Illnesses in the Indoor Environment. *In* Unhealthy Housing: Research, Remedies and Reform, edited by R. Burridge and D. Ormandy. London, E&FN Spon, pp 117-140.

COLLINS, K.J. EASTON, J.C. BELFIELD-SMITH, H. EXTON,-SMITH, A.N. and PLUCK, R.A. (1985) Effects of age on body temperature and blood pressure in cold environments. *Clinical Science*, vol 69, pp.465-70.

COMBINED HEAT AND POWER GROUP (1977) District Heating Combined with Electricity Generation in the UK. Department of Energy, Energy Paper 20, London, HMSO.

COMBINED HEAT AND POWER GROUP (1979a) Heat Loads in British Cities. Department of Energy, Energy Paper 34, London, HMSO.

COMBINED HEAT AND POWER GROUP (1979b) Combined Heat and Electricity Power Generation in the UK: Report to the Secretary of State for Energy. Department of Energy, Energy Paper 35, London, HMSO.

COMMISSION OF THE EUROPEAN COMMUNITIES (1990) Green Paper on the Urban Environment. Brussels, CEC.

COMMISSION OF THE EUROPEAN COMMUNITIES. (1992) Towards Sustainability. Brussels, CEC.

CONNAUGHTON, J.N. and MUSANNIF, A.A.B. (1990) Low Energy Housing at Little or No Additional Cost: Report of a Feasibility Study. Energy Efficiency Office, RD 53/42.

COOK, S.W. and BERRENBERG, J.L. (1981) Approaches to Encouraging Energy Conservation Behaviour: A Review and Conceptual Framework. *Journal of Social Issues*, vol 37 (2).

CRAIG, C.S. and McCANN, J.M. (1978) Assessing Communication Effects on Energy Conservation. *Journal of Consumer Research*, vol 5, pp 82-88.

CRAWSHAW, A.J.E. WILLIAMS, D.I. and CRAWSHAW, C.M. (1985) Consumer Knowledge and Electricity Consumption. *Journal of Consumer Studies and Home Economics*, vol 9, pp 283-289.

CULLINGWORTH J B. AND SPARLING W J. (1988) Community Energy Planning: Projects and Potentialities. *In* Planning for Changing Energy Conditions. edited by J Byrne and D Rich, New Brunswick, N.J. Transaction Books.

DANISH MINISTRY OF ENERGY (1987) District Heating. Research and Technological Development in Denmark. Copenhagen, Danish Energy Agency.

DANISH MINISTRY OF ENERGY (1990) Energy 2000. Copenhagen, Danish Energy Agency.

DANISH MINISTRY OF ENERGY (1993) District Heating. Research and Technological Development in Denmark. Copenhagen, Danish Energy Agency.

DARLEY, J.M. and BENIGER, J.R. (1981) Diffusion of Energy-Conserving Innovations. *Journal of Social Issues*, vol 37 (2), pp 150-171.

DAVIES, H. and PYLE, J. (1993) Influences on the Construction of Energy Efficient Housing. *In* A Focus for Building Surveying Research. *In* Proceedings of a seminar held at the Royal Institution of Chartered Surveyors, London, RICS.

DEn (1989) An Evaluation of Energy Related Greenhouse Gas Emissions and Measures to Ameliorate Them. Department of Energy, Energy Paper 58, HMSO, London.

DEUTSCHER BUNDESTAG (1991) Protecting the Earth: A Status Report with Recommendations for a New Energy Policy. Third Report of the Enquete Commission on Preventative Measures to Protect the Earth's Atmosphere. Bonn.

DILLMAN, D.A. ROSA, E.A. DILLMAN, J.J (1983) Lifestyle and Home Energy Conservation in the U.S. *Journal of Economic Psychology*, vol 3, pp 299-315 (cited in Lutzenhiser, 1993).

DOE & WELSH OFFICE (1994) Building Regulations Approved Document, Part L: Conservation of Fuel and Power. London, HMSO.

DOE (1981). Energy in Housing I. The Better Insulated House Programme. Department of the Environment, London, HMSO.

DOE (1991) English House Condition Survey: 1986 - Supplementary (energy) Report. London, HMSO.

DOE (1991). Housing and Construction Statistics 1981 to 1990. Central Statistical Office, London, HMSO.

DOE (1992) Climate Change: Our National Programme for CO_2 Emissions. London, Department of Environment, Global Atmosphere Division.

DOE (1993a) English House Condition Survey: 1991. Department of Environment, London, HMSO.

DOE (1993b) Green House First Report 1993. London, Department of the Environment.

DOE (1993c) Environmental Appraisal of Development Plans. Department of Environment, London, HMSO.

DOE (1994a) Digest of Environmental Protection and Water Statistics 1994, London, HMSO.

DOE (1994b) Energy Efficiency in Council Housing: Exemplary Project Guide: Green House Programme. London, Department of the Environment.

DOE (1994c) The York Experience. Green House Profile no. 4, Produced for the Department of the Environment by the Building Research Energy Conservation Support Unit, Watford, UK.

DOE (1994d) Energy Efficiency in Council Housing: Guidance for Local Authorities. London, Department of the Environment.

DOE et al (1990) This Common Inheritance: Britain's Environmental Strategy. London, HMSO.

DOE et al (1994a) The UK Programme on Climate Change. Department of Environment, London, HMSO.

DOE et al (1994b) Sustainable Development: The UK Strategy. Department of Environment, London, HMSO.

DTI (1992) Energy Related Carbon Emissions in Possible Future Scenarios for the United Kingdom. Department of Trade and Industry, Energy Paper 59. London, HMSO.

DTI (1993) Digest of United Kingdom Energy Statistics 1993. Department of Trade and Industry, London, HMSO.

DUPONT, P. and MORRILL, J. (1989) Residential Indoor Air Quality & Energy Efficiency. USA, American Council for an Energy-Efficient Economy, Series on Energy Conservation and Energy Policy.

DUMONT, R.S. (1992) Advanced Houses Program: Technical Requirements. Ottawa, Ontario, Canadian Centre for Mineral and Energy Technology (CANMET), Buildings Group.

DUMONT, R.S. (1993) A Short History of Low Energy Houses for Cold Climates, With an Emphasis on Canadian Contributions. *In* Proceedings Innovative Housing Conference, Vancouver, Canada.

DUNSTER J.E. (1994c) Domestic Energy Fact File. Volume 3 Private Rented Homes. BR 273, Building Research Establishment, Watford, UK.

DUNSTER J.E. (1994a) Domestic Energy Fact File. Volume 1 Owner Occupied Homes. BR 271, Building Research Establishment, Watford, UK.

DUNSTER J.E. (1994b) Domestic Energy Fact File. Volume 2 Local Authority Homes. BR 272, Building Research Establishment, Watford, UK.

EC (1994) Council Directive 94/2/EC implementing Council Directive 92/75/EEC with regard to energy labelling of household electric refrigerators, freezers and their combinations. *Official Journal of the European Communities*, vol 37, L 45/1-2.

ECLIPSE RESEARCH (1993a) Organisational Aspects of Energy Management. General Information Report 12. Building Research Establishment Energy Conservation Support Unit, Watford, UK.

ECLIPSE RESEARCH (1993b) Reviewing Energy Management. General Information Report 13, Building Research Establishment Energy Conservation Support Unit, Energy Efficiency Office, Watford, UK.

EDGETECH IG (1990-) Edgetech Newsletter. Ottawa, Canada, Edgetech IG Ltd.

EDWARDS, N. (1990) Introducing the Thick Building. *Building Services*, vol 12 (1).

EEC (1993) Council Directive 91/76/EEC of 13 september 1993 to limit carbon dioxide emissions by improving energy efficiency (submitted in the framework of the SAVE programme). *Official Journal of the European Communities,* Ref. L237, 22 September 1993.

EEO (1991) Insulating Your Home. Information leaflet prepared by the Energy Efficiency Office for the 'Helping the Earth Begins at Home' Campaign. London, HMSO.

EEO/W.S. ATKINS & PARTNERS (1986) Heating Homes and Households: Monitoring and Evaluation Report, Willows Energy Efficiency Advice Programme. London, Energy Efficiency Office, Department of the Environment (cited in Salvage 1992).

EKINS, P. and COOPER, I. (1993) Cities and Sustainability. Clean Technology Unit, AFRC-SERC, Swindon, UK.

ELKIN, T. and McLAREN, D. (1991) Reviving the City. London, Friends of the Earth.

ENERGY RESEARCH AND DEVELOPMENT GROUP. (1980) Energy Efficient Housing: A Prairie Approach. Alberta Energy and Natural Resources Energy Conservation Branch, Saskatchewan Mineral Resources Office of Energy Conservation, Manitoba Energy & Mines Conservation and Renewable Energy Branch.

ENERGY SAVING TRUST (undated) Promoting Energy Efficiency and Conservation. publicity material, London, Energy Saving Trust.

ENVIRONMENT CITY (1993) Stepping Stones. Lincoln, Rural Society for Nature Conservation.

EVANS, B. (1990) Build it Green: Superinsulation. *Architects Journal*, 7 March.

EVANS, R.D. (1990) Environmental and Economic Implications of Small Scale CHP. Energy & Environment Paper No. 3, Department of Energy, London.

EVANS, R.D. and HERRING, H.P.J. (1989) Energy Use and Energy Efficiency in the UK Domestic Sector up to the Year 2010. Energy Efficiency Series 11, Department of Energy, London.

EVERETT, R. & ANDREWS, D. (1986) An Introduction to Domestic Micro-CHP. Open University, ERG-056, Milton Keynes, UK.

EVERETT, R. (1992) Domestic Micro-CHP and the Greenhouse Effect. Open University, EERU-069, Milton Keynes, UK.

EVERETT, R. HORTON, A. DOGGART, J. with WILLOUGHBY, J. (1985) Linford Low Energy Houses. Open University ERG 050/ETSU-S-1025, Milton Keynes, UK.

FARHAR, B.C. (1993) Trends in Public Perceptions and Preferences on Energy and Environmental Policy. NREL/TP 461-485. Washington, DC: National Renewable Energy Laboratory, (cited in Lutzenhiser, 1993).

FEIST, W. JÄKEL, M. STURM, H.J. WERNER, J. (1994) Lüftung im Passivhaus. Passivhaus-bericht Nr 8, Germany, Darmstadt, Institut Wohnen und Umwelt.

FINBOW, M. and PICKERING, M. (1988) Passive Solar Housing Layout Study. Department of Energy, ETSU S 1126.

FISK, D.J. (1978) Microeconomics and the Demand for Space Heating. BRE CP 6/78, Building Research Establishment, Watford, UK.

GASKELL, G. and ELLIS, P. (1981) Energy Conservation: A Psychological Perspective on a Multidisciplinary Phenomenon. *In* Confronting Social Issues: Some Applications of Social Psychology - Volume 1. edited by P. Stringer. London, Academic Press, pp 103-122.

GASKELL, G. and PIKE, R. (1983) Residential Energy Use: An Investigation of Consumers and Conservation Strategies. *Journal of Consumer Policy*, vol 6, pp 285-302.

GELLER, E.S. WINETT, R.A. and EVERETT, P.B. (1982) Preserving the Environment: New Strategies for Behaviour Change. New York, Pergamon.

GILLETT, W.B. & STAMMERS, J.R. (1992) Review of Active Solar Technologies. Department of Trade and Industry, ETSU S 1337.

GLASSON J. (1995) Regional Planning and the Environment: Time for a SEA Change?, *Urban Studies*, vol 27 (8) pp 769-776.

GOODCHILD, B. (1994) Housing Design, Urban Form and Sustainable Development. *Town Planning Review*, vol 65 (2).

GORDON, H.G. MCDOUGAL, J.R. RITCHIE, J.R.B. and CLAXTON, J.D. (1981) Analysis of Consumer Energy Consumption. *In* Consumers and Energy Conservation: International Perspectives on Research and Policy Options, edited by J.D. Claxton et al. New York, USA, Prager Publishers.

GREEN, H. and VENTRIS, N. (1983) Attitudes to Energy Use in a Housing Action Area. *In* Consumers, Buildings and Energy, edited by B. Stafford. Birmingham, Centre for Urban & Regional Studies, University of Birmingham.

GREEN, J. (1992) The Great Gonerby Energy Efficiency Initiative: A Local Project to Reduce Electricity Demand Through Energy Efficiency Improvements. Newcastle-upon-Tyne, Neighbourhood Energy Action.

GREEN, J. (1994) Evaluation of the Great Gonerby Energy Efficiency Initiative. Newcastle-upon-Tyne, Neighbourhood Energy Action.

HACKETT, B. and LUTZENHISER, L. (1991) Social Structures and Economic Conduct: Interpreting Variations in Household Energy Consumption. *Sociological Forum*, vol 6 (3), pp 449-70.

HALCROW GILBERT ASSOCIATES (1992) Fenestration 2000 - Phase II: Review of Advanced Glazing Technology and Study of Benefits for the UK. Department of Energy, ETSU S1342.

HANSEN, J. et al (1981) Climate Impact of Increasing Atmospheric Carbon Dioxide. *Science*, vol 21 (3), pp 957-966.

HEALY, P. and SHAW, T. (1993) Planners, Plans and Sustainable Development. *Regional Studies,* vol 27 (8), pp 769-772.

HEBBERT, M. (1992) Environmental Foundation for a New Kind of Town and Country Planning. *Town and Country Planning*, vol 61 (6).

HEDGES, A. (1991) Attitudes to Energy Conservation in the Home: Report on a Qualitative Study. London, HMSO.

HENDERSON, G. and SHORROCK, L. (1989) Cutting off the gases. *Building Services*, vol 11 (10).

HENDERSON, G. and SHORROCK, L.D. (1990) Greenhouse Gas Emissions and Buildings in the United Kingdom. BRE IP2/90, Building Research Establishment, Watford, UK.

HESKETH, J.L. (1975) Fuel Debts: Social Problems in Centrally Heated Council Housing. Manchester, Family Welfare Association (cited in Parker, 1983).

HILL, R. PEARSALL, N.M. and CLAIDEN, P. (1992) The Potential Generating Capacity of PV-Clad Buildings in the UK: Volume 1. Department of Trade & Industry, ETSU S 1365-P1.

HOUGHTON, J. T. JENKINS, G.J. and EPHRAUMS, J.J. (eds.) (1990) Climate Change - the IPCC Scientific Assessment. Cambridge Univsrity Press.

HOUGHTON J.T. CALENDER B.A. and VARNEY S.K. (eds.) (1992) Climate Change 1992 - the Supplementary Report to the IPCC Scientific Assessment. Cambridge University Press.

HUNT, S. M. and BOARDMAN, B. (1994) Defining the problem. *In* Domestic Energy and Affordable Warmth. Edited by T.A Markus. The Watt Committee on Energy, Report no.30. E&FN Spon, London.

HUNT, S.M. (1993) Damp and Mouldy Housing: A Holistic Approach. *In* Unhealthy Housing: Research, Remedies and Reform. edited by R. Burridge and D. Ormandy. London, E&FN Spon, pp 69-93.

HUNT, S.M. MARTIN, C.J. PLATT, S. LEWIS, C. and MORRIS, G. (1988) Damp Housing Mould Growth and Health Status: Part I, Report to the Funding Bodies, Glasgow and Edinburgh District Councils. Health and Behavioural Change Research Unit, University of Edinburgh.

HUNT, S.M. and LEWIS, C. (1988) Damp Housing, Mould Growth and Health Status: Part II, Report to the Funding Bodies, Glasgow and Edinburgh District Councils. Health and Behavioural Change Research Unit, University of Edinburgh (cited in Hunt 1993).

HUTCHINSON, D. (1990) Combined Heat and Power. *In* Planning for Sustainable Development. London, Town and Country Planning Association.

HUTTON, S. GASKELL, G. PIKE, R. BRADSHAW, J. and CORDON, A. (1985) Energy Efficiency in Low Income Households: An Evaluation of Local Insulation Projects. London, HMSO.

ILLICH, I.D. (1974) Energy and Equity. Calder and Boyars, London, UK.

JESCH, J.F. POLLOCK, A. and JANKOVIC, L. (1987) The Householder as Energy Manager: A Survey of User Attitudes in Low Energy Housing. *In* Proceedings of the Third International Congress on Building Energy Management, Lausanne, Switzerland, September 28-October 2, pp 403-410.

KEATING M. (1993) Agenda for Change. Geneva, Centre for our Common Future.

KEATINGE, W.R. (1986) Seasonal Mortality Among Elderly People with Unrestricted Home Heating. *British Medical Journal*, vol 293, pp 732-733.

KEATINGE, W.R. COLESHAW, S.K.R. COTTER, F. MATTOCK, M. MURPHY, M. and CHELLIAH, R. (1984) Increases in platelet and red cell counts, blood viscosity and arterial pressure during mild surface cooling: factors in mortality from coronary and cerebral thrombosis in winter. *British Medical Journal,* vol. 289, 24 November 1984.

KEATINGE, W.R. COLESHAW, S.R.K and HOLMES, J. (1989) Changes in Seasonal Mortalities with Improvement in Home Heating in England and Wales from 1964 to 1984. *International Journal of Biometeorology*, vol 33, pp 71-76.

KEMPTON, W. and MONTGOMERY, L. (1982) Folk Quantification of Energy. *Energy*, vol 7 (10), pp 817-827.

KEPLINGER, D. (1978) Site Design and Orientation for Energy Conservation. *Ekistics*, No. 269.

KOEHLER, N. (1987) Energy Consumption and Pollution of Building Construction. *In* Proceedings of International Congress on Building Energy Management, Lausanne.

KORSGAARD, J. (1979) The effect of the indoor environment on the house dust mite. *In* Indoor Climate: Effects on Human Comfort, Performance and Health, Edited by P.O. Fanger, and O. Valbjorn. Danish Building Research Institute, Copenhagen. (Cited in Hunt 1993)

KRAUSE, F. et al. (1989) Energy Policy in the Greenhouse. Final Report of the International Project for Sustainable Energy Paths (IPSEP), El Cerrito, California, USA.

KREISI, R. (1989) Null-Heizenergie Konzept in einer Siedlung in Wädenswil. *Schweizer Ingenieur und Architekt*, No 45, 9th November 1989.

LAMPERT, C.M. & MA, Y.P. (1992) Fenestration 2000 - Phase III: Advanced Glazing Materials Study. Department of Trade and Industry, ETSU S1215.

LEACH, G. (1981) Energy-Related Statistics for UK Dwellings. London, International Institute for Environment and Development.

LEACH, G. et al. (1979) A Low Energy Strategy for the UK. Science Reviews

LIDDAMENT, M.W. (1993) Energy Efficient Ventilation Strategies. *In* Proceedings Innovative Housing Conference, Vancouver, Canada.

LOCKE, D. (1994) Sustainable Patterns of Development. *Town and Country Planning*, vol 63 (9).

LONDON BUSINESS SCHOOL (1988) The Marketability of Passive Solar House Designs: Phase One. Department of Energy, ETSU S 1128A.

LORENZ, E. et al. (1963) Deterministic, Nonperiodic Flow. *Journal of the Atmospheric Sciences*, vol 20, pp 448-464.

LOVINS, A.B. (1977) Soft Energy Paths: Toward a Durable Peace. UK, Penguin.

LOVINS, A.B. (1981) Energy Strategy for Low Climatic Risks. Report under contract 104 02 513 to Federal Environmental Agency, West Germany, June 1981.

LOWE R.J. (1994) Controlling CO_2 Emissions - the Fiscal Option. *In* Proceedings of World Renewal Energy Congress, Reading, UK.

LOWE, R.J. CHAPMAN, J. and EVERETT, R.C. (1985) The Pennyland Project. Open University Energy Research Group, ERG 053/ETSU-S-1046, Milton Keynes, UK.

LOWE, R.J., CURWELL S.R., BELL, M. and AHMAD, A. (1994) Airtightness in Masonry Dwellings: Laboratory and Field Experience. *Building Services Engineering Research & Technology*, 15 (3).

LUSSER, M. (1994) Environmental Planning and Sustainable Development. *Town and Country Planning*, vol 63 (6).

LUTZENHISER, L. (1992) A Cultural Model of Household Energy Consumption. *Energy*, vol 17 (1), pp 47-60.

LUTZENHISER, L. (1993) Social and Behavioural Aspects of Energy Use. *Annual Review of Energy and the Environment*, vol 18, pp 247-289.

MACNEILL, J. WINSEMIUS, P. AND YAKUSHIJI, T. (1991) Beyond Interdependence. New York, Oxford University Press.

MANT, D.C. and GRAY, J.A. (1986) Building Regulation and Health. Building Research Establishment Report, Watford, UK, and Department of the Environment.

MARCH CONSULTING GROUP. (1990) Energy Efficiency in Domestic Electricity Appliances. Department of Energy, Energy Efficiency Series No 13.

MARKUS, T.A. (1993) Cold, Condensation and Housing Poverty. *In* Unhealthy Housing: Research, Remedies and Reform. edited by R. Burridge and D. Ormandy. London, E&FN Spon, pp 141-167.

MARKUS, T.A. and MORRIS, E.N. (1980) Buildings, Climate and Energy. Pitman, London.

MAYERS, H. (1993) Findings of Energy Audit and Questionnaire Survey into Heating of Local Authority Households. Report to the CHILL Campaign and the London Borough of Greenwich, London, Tenants Energy Advice Service.

McINTYRE D.A. (1986) Domestic Ventilation Heat Recovery Using Heat Pumps. Electricity Council Research Centre, ECRC/M2065, September 1986.

MEYER, N.I. & NORGÅRD, J.S. (1989) Planning Implications of Electricity Conservation: The Case of Denmark. *in* Electricity: Efficient End-use and New Generation Technologies and their Planning Implications. Johansson, T.B. et al (eds.), Lund University Press, 1989.

MILLER, R.D. and FORD, J.M. (1985) Shared Savings in the Residential Market: A Public/Private Partnership for Energy Conservation. Baltimore, MD: Energy Task Force, Urban Consortium for Technological Initiatives. USA.

NÄNNI, J. BRUNNER, C.U. SPÖRRI, R. and ZÜRCHER, C. (1989) Rationelle, rechnerunterstützte Bauteilanalyse bei Wärmebrücken. *Bauphysik*, vol 11.

NEA (1992) Home Energy Efficiency Scheme: The First Fifteen Months. Newcastle-upon-Tyne, Neighbourhood Energy Action.

NEA (1993) Home Energy Efficiency Scheme: Barriers to Access. Newcastle-upon-Tyne, Neighbourhood Energy Action.

NEA (1994a) Fuel Poverty Research. *In* Energy Action, July 1994 No. 58 p 6, Newcastle-upon-Tyne, Neighbourhood Energy Action.

NEA (1994b) Energy Efficiency Measures: Assessing and Improving Social Housing. Newcastle-upon-Tyne, Neighbourhood Energy Action.

NORGÅRD, J.S. (1979) Improved Efficiency in Domestic Electricity Use. *Energy Policy*, vol 7, March 1979.

NORGÅRD, J.S. (1989) Low Electricity Appliances: Options for the Future. *In* Electricity: Efficient End-use and New Generation Technologies and their Planning Implications. Johansson, T.B. et al (eds.), Lund University Press, 1989.

NORGÅRD, J.S. & CHRISTENSEN, B. (1995) Technological Options and Sustainable Energy Welfare. *In* Samuels & Prasad (eds.) (1995).

NISSON, J.D.N. (1988) Residential Building Design and Construction Workbook. 2nd edn. Arlington, MA USA, Cutter Information Corporation.

NITSCHKE, I.A. et al. (1985) Indoor Air Quality, Infiltration and Ventilation in Residential Buildings. NYSERDA report No. 85-10. Albany, NY, USA.: New York State Energy Research and Development Authority. (Cited in Dupont and Morrill, 1989).

NORTH WEST REGIONAL ASSOCIATION. (1994) Greener Growth. UK, Wigan, North West Regional Association.

NORTON, B. and LOCKHART-BALL, H. (1989) Daylighting Buildings. *In* Proceedings of ISES Conference, Imperial College, London, April 1989.

OECD (1991) Environmental Policy for Cities in the 1990's. Paris, OECD.

OLIVIER, D. (1989) Beyond the Building Regulations. *Building Services*, vol 11 (5).

OLIVIER, D. (1991) Continental Efficiency. *Building Services*, vol 13 (3).

OLIVIER, D. (1992) Energy Efficiency and Renewables: Recent Experience on Mainland Europe. UK, Energy Advisory Associates.

OLIVIER, D. (1994) A Low Watts House. *Building Services*, vol 16 (7).

OLSEN, M.E. (1981) Consumer's Attitudes Toward Energy Conservation. *Journal of Social Issues*, vol 37 (2), pp 108-131.

OPTIMA ENERGY SERVICES (1989) Neighbourhood Energy Advisers. Final report of a research project undertaken by Optima Energy Services in association with British Gas, Optima Energy Services.

OSELAND, N.A. and WARD, D.D. (1993) An Interim Evaluation of the Home Energy Efficiency Scheme (HEES). Building Research Establishment Client Report CR62/93, Watford, UK.

OWENS, S. (1986) Energy, Planning and Urban Form, Pion, London.

OWENS, S. (1992) Energy, Environmental Sustainability and Land-use Planning. *In* Sustainable Development and Urban Form. edited by M. Breheny. Pion, London.

PALIN, S.L. WINSTANLEY, R. MCINTYRE, D.A. and EDWARDS, R.E. (1993) Energy Implications of Domestic Ventilation Strategy. *In* Proceedings 14th AIVC Conference Copenhagen, Denmark.

PARKER, G. (1983) Debt. *In* Energy and Social Policy. edited by J. Bradshaw and T. Harris, London, Routledge & Kegan Paul, pp 58-68.

PERERA, E. and PARKINS, L. (1992) Airtightness of UK Buildings: Status and Future Possibilities. *Environmental Policy and Practice*, vol 2 (2).

PERSSON, A. (1994) Well-insulated and Attractive Windows. *CADDET Energy Efficiency Newsletter*, 1994 (2).

POLICH, M.D. (1984) Minnesota RCS: The Myths and Realities. *In* Doing Better: Setting an Agenda for the Second Decade, Vol. G. American Council for an Energy-Efficient Economy. Washington. USA.

POTTER S. AND HUGHES P. (1990) Vital Travel Statistics. London, Transport 2000.

PRESIDENT'S COUNCIL ON ENVIRONMENTAL QUALITY & DEPARTMENT OF STATE. (1980) The Global 2000 Report to the President: Entering the Twenty-first Century. Washington DC, USA, US Government Printing Office.

PRIOR, J.J. & BARTLETT, P.B. (1995) Environmental Standard: Homes for a Greener World. BR278, Building Research Establishment, Watford, UK.

PYLE, J. (1994) Personal Communication.

PYLE, J. and DAVIES, H. (1994) Energy Efficient Housing: Pay Now and Save Later. *Architects and Surveyors Institute Journal*, vol 5 (5), pp 7-8.

QUIGLEY, J.M. (1986) Comment: "Blind Spots" in Perspective. *Journal of Policy Analysis and Management*, vol 5 (2), John Wiley.

RAVETZ, J. CARTER, G. FOX, J. ROBERTS, P. ROOKWOOD, R. SERVANTE, D. WINTER, P. AND YOUNG, S. (1995) Manchester 2020: Sustainable Development in the City Region. London Town and Country Planning Association.

RAVETZ, J. (1994) Manchester 2020: A Sustainable City Region Demonstration Project. London, Town and Country Planning Association.

RAW, G.J. (1988) A Discussion Paper: Health and Hygrothermal Conditions in Homes. *In* Planning, Physics and Climate Technology for Healthier Buildings, Vol 2. Proceedings of the Healthy Buildings 88, Conference, Stockholm, Sweden.

RICHARDSON, P. (1978) Fuel Poverty, Papers in Community Studies No. 20, Department of Social Administration and Social Work, University of York (cited in Parker, 1983).

RICS (1994) Financial Incentives for Greener Homes. London, Royal Institution of Chartered Surveyors.

RMDP/NEDO (1989) Home Automation: Will the Public Buy It ? Consumer Attitudes to Home Automation. Brighton, RMDP Ltd. and National Economic Development Office.

ROBERTS, P. (1995) Environmentally Sustainable Business: A Local and Regional Perspective, London, Paul Chapman Publishing.

ROBERTS, P. (1994) Sustainable Regional Development. *Regional Studies*, vol 28 (8).

ROBERTSON, I.T. and SMITH, M. (1985) Motivation and Job Design. London, Institute of Personnel Management.

ROBINSON, O. and LITTLER, J. (1991) Courtyard Passive Solar Houses Demonstration Project - Summary Report. CEC.

ROSENFELD, A. and LYNN, P. (1993) Energy Efficient Technologies and Policies Can Help Us Win the Race to Save the Planet. *In* Proceedings Innovative Housing Conference, Vancouver Canada.

ROSS, A. COLLINS, M. and SANDERS, C. (1990) Upper Respiratory Tract Infection in Children, Domestic Temperatures, and Humidity. *Journal of Epidemiology and Community Health*, vol 44, pp 142-146.

ROYAL COMMISSION ON ENVIRONMENTAL POLLUTION. (1994) Transport and the Environment. (Cm 267), London, HMSO.

RUYSSEVELT, P. & MARTIN, C. (1987) A Structured Series of Test Room Experiments in the UK. *In* Proceedings of European Conference on Architecture, Munich, April 1987.

RUYSSEVELT, P. LITTLER, J. and CLEGG, P. (1987) Superinsulated Houses for the UK Using a Fully Pre-fabricated Timber Frame. *In* Proceedings of European Conference on Architecture, Munich, April 1987.

RYDIN Y. (1992) Environmental Dimensions of Residential Development and the Implications for Local Planning Practice, *Journal of Environmental Planning and Management*, vol 35 (1), pp 43-61.

SALVAGE, A.V. (1992) Energy Wise? Elderly People and Domestic Energy Efficiency. Age Concern Institute of Gerontology, King's College, London.

SALVAGE, A.V. (1993) Elderly People in Cold Homes: The Implications for Energy Efficiency and Housing. Housing Review, vol 42 (5), pp 78-80.

SALVAGE, A.V. (1994) Heating Controls for Elderly People: Report on an Exploratory Research Project. London, Age Concern Institute of Gerontology, King's College London.

SAMUELS, R. & PRASAD, D.K. eds. (1995) Global Warming and the Built Environment. Spon, London.

SANDERS, M.S. and McCORMICK, E.J. (1993) Human Factors in Engineering and Design. 7th. edition. McGraw-Hill International.

SCHIPPER, L. and KETOFF, A. (1985) Residential Energy Use in the OECD. *Energy Journal*, vol 6, pp 65-85. (cited in Lutzenhiser, 1992).

SCHIPPER, L. KETOFF, A. and KAHANE, A. (1985) Explaining Residential Energy Use by International Bottom-up Comparisons. *Annual Review of Energy*, vol 10, pp 341-405. (cited in Lutzenhiser, 1992).

162

SCHOLES, W.E. and FOTHERGILL, L.C. (1975) Energy Conservation: A Study of Energy Consumption in Buildings and Possible Means of Saving Energy in Housing. CP56/75, Building Research Establishment, Watford, UK.

SCHULTZ, J.M. (1992) Ramme-/karmkonstructioner til højisolerende vinduer. Thermal Insulation Laboratory, Report No. 237, Technical University of Denmark, Lyngby.

SCIVYER, C. PERERA, E. and WEBB, B. (1994) Build Tight: the Orkney Experience. *Building Services*, vol 16 (7).

SEABOURN, M (1995) Putting councils in the conservation hot seat. *Inside Housing*, 18 August 1995.

SELIGMAN, C. and DARLEY, J.M. (1977) Feedback as a Means of Decreasing Residential Energy Consumption. *Journal of Applied Psychology*, vol 62 (4), pp 363-368.

SELIGMAN, C. DARLEY, J.M. and BECKER, L. (1978) Behavioural Approaches to Residential Energy Conservation. *Energy and Buildings*, vol 1, pp 325-337.

SELIGMAN, C. KRISS, M. DARLEY, J.M. FAZIO, R.H. BECKER, L.J. and PRYOR, J.B. (1979) Predicting Residential Energy Consumption from Homeowner's Attitudes. *Journal of Applied Social Psychology*, vol 9, pp 70-90 (cited in Olsen, 1981).

SHORROCK, L D. and BROWN, J H F. (1993) Domestic Energy Fact File 1993 Update. BR251, Building Research Establishment, Watford, UK.

SHORROCK, L.D. and HENDERSON, G. (1990) Energy Use in Buildings and Carbon Dioxide Emissions. Building Research Establishment, Watford, UK.

SHORROCK, L.D. HENDERSON, G. and BROWN, J.H.F. (1992) Domestic Energy Fact File. Building Research Establishment Report, Watford, UK.

SINHA, R. and MAYO, T. (1993) Emerging Trends in the Advanced Houses Program. *In* Proceedings Innovative Housing Conference, Vancouver, Canada.

SIVIOUR, J.B. (1994) Experimental U-values of Some House Walls. *Building Services Engineering Research and Technology*, vol 15 (1).

SONDEREGGER, R.C. (1978) Movers and Stayers: The Resident's Contribution to Variation Across Houses in Energy Consumption for Space Heating. *In* Saving Energy in the Home: Princeton's Experiments at Twin Rivers. edited by R.H Socolow. Cambridge, Mass. USA, Ballinger.

STERN, P.C. (1986) Blind Spots in Energy Analysis: What Economics Doesn't Say About Energy Use. *Journal of Policy Analysis and Management*, vol 5 (10) pp 200-227. (cited in Lutzenhiser 1992).

STERN, P.C. and ARONSON, E. (1984) Energy Use: The Human Dimension. National Research Council, W. H. Freeman and Company, New York.

STERN, P.C. BLACK, J.S. and ELWORTH, J.T. (1981)   Home Energy Conservation: Issues and Programmes for the 1980s. Mount Vernon, New York, Consumers Union Foundation, (cited in Stern and Aronson, 1984).

STERN, P.C. BLACK, J.S. and ELWORTH, J.T. (1982)   Influences on Household Energy Adaptations. Paper presented to the American Association for the Advancement of Science, Washington D.C.

STRACHAN, D.P. (1993) Dampness, Mould Growth and Respiratory Disease in Children. *In* Unhealthy Housing: Research, Remedies and Reform. edited by R. Burridge and D. Ormandy. London, E&FN Spon, pp 94-116.

TAYLOR, L. (1992) Employment Aspects of Energy Efficiency. *In* Energy Efficiency: The Policy Agenda for the 1990s. Christie, I. and Ritchie, N. (eds.) London, Policy Studies Institute.

TODD, R.W. and ALTY, C.T.N. (1977) An Alternative Energy Strategy for the UK. 1st ed., National Centre for Alternative Technology, Macchynlleth, UK.

TONG, D. (1987)   Energy Efficiency Through Insulation and Education. *In* Energy Action, Newcastle Upon Tyne, Neighbourhood Energy Action, pp 16-17.

TONN, B.E. and BERRY, L. (1985)   An Illustration of Household Energy-Conservation Modelling Using the Search-List Decision-Heuristic. *Environment and Planning A*, vol 17 pp 1243-1261.

U.K. GOVERNMENT. (1994) Sustainable Development: The UK Strategy. (Cm 2426), London, HMSO.

UN (1992) Conference on Environment & Development, Rio de Janiero, 1992. United Nations, New York, USA (UN Information Centre, London, UK.)

UN (1993a) Agenda 21 Earth Summit: The UN Programme of Action from Rio. United Nations, New York, USA (UN Information Centre, London, UK.)

UN (1993b) Earth Summit Framework Convention on Climate Change. United Nations, New York, USA (UN Information Centre, London, UK.)

VALE, B. and VALE, R. (1993)   Building the Sustainable Environment. *In* Planning for a Sustainable Environment. edited by A. Blowers, Earthscan, London.

VERHALLEN, T.M.M. and VAN RAAIJ, F.W. (1981)   Household Behaviour and the Use of Natural gas for Home Heating. *Journal of Consumer Research*, vol 8.

WARD, D.D. (1991) The Temperature Benefit from Domestic Energy Efficiency Measures. BRE CR114/91, Building Research Establishment, Watford, UK.

WATT COMMITTEE ON ENERGY (1979) Energy Development and Land in the United Kingdom. London, Watt Committee on Energy.

WEBSTER, P.J. (1987) The Salford Low Energy House: A Demonstration at Strawberry Hill, Salford.   Salford University Civil Engineering Dept. for

Building Research Establishment Energy Conservation Support Unit. Report ED 179/59.

WEBZELL, A.P. and FEWINGS, P.T. (1994) Comparative Cost Analysis of an Existing House with the Same House Constructed of Environmentally Friendly Materials. *In* Buildings and the Environment, Proceedings of the First International Conference. Watford, Building Research Establishment, CIB.

WICKS, M. (1978) Old and Cold: Hypothermia and Social Policy. Heinemann Educational Books (cited in Wicks, 1983).

WICKS, M. (1983) Cold Conditions, Hypothermia and Health. *In* Energy and Social Policy. edited by J. Bradshaw and T. Harris, London, Routledge & Kegan Paul, pp 85-102.

WILLIAMS, D.I. and CRAWSHAW, C.M. (1986) Advanced Control Systems and the Householder. *In* Contemporary Ergonomics 1986, Proceedings of the Ergonomics Society's 1986 Annual Conference Durham, UK, 8-11 April 1986. edited by D.J. Osborne. London, Taylor Francis Limited.

WILLIAMS, D.I. and CRAWSHAW, C.M. (1987) Cognitive Factors in Domestic Energy Management. *In* Proceedings of International Conference on Building Energy Management, Lausanne.

WILLIAMS, D.I. CRAWSHAW, A.J.E. and CRAWSHAW, C.M. (1985) Energy Efficiency and the Domestic Consumer. *The Journal of Interdisciplinary Economics*, vol 1, pp 19-27.

WILLIAMS, R.H. (1989) Innovative Approaches to Marketing Electric Efficiency. *In* Electricity: Efficient End-Use and New Generation Technologies, and Their Planning Implications. T.B Johansson et al. (eds.), Lund University Press.

WINCH, D.M. (1971) Analytical Welfare Economics. Middlesex, Penguin Books.

WINETT, R.A. KAISER, S. and HABERKORN, G. (1977) The Effects of Monetary Rebates and Daily Feedback on Electricity Conservation. *Journal of Environmental Systems*, vol 6 (4), pp 329-341.

WINETT, R.A. LOVE, S.Q. and CHARLOTTE, K. (1982) The Effectiveness of an Energy Specialist and Extension Agents in Promoting Summer Energy Conservation by Home Visits. *Journal of Environmental Systems*, vol 12 (1), pp 61-70.

WORLD COMMISSION ON ENVIRONMENT AND DEVELOPMENT. (1987) Our Common Future. Oxford, Oxford University Press.

WORLD HEALTH ORGANISATION (1984) The Effects of the Indoor Housing Climate on the Health of the Elderly. World Health Organisation, Copenhagen

XU, S. and MADDEN, M. (1989) Urban Ecosystems: A Holistic Approach to Urban Analyses and Planning. *Environment and Planning B*, vol 16.

YANNAS, S. (1994) Solar Energy and Housing Design. (2 vols.) London, Architectural Association.

Acts of Parliament

Home Energy Conservation Act 1995. London, HMSO.

Index

167